Band 22 · Februar 1986 **2**

isotopenpraxis

AKADEMIE-VERLAG BERLIN

ISSN 0021-1915 Isotopenpraxis, Berlin 22 (1986) 2, 41–72

An unsere Autoren

Die Zeitschrift veröffentlicht Beiträge zu folgenden Themen: Kontrolle und Analyse radioaktiver und stabiler Isotope, Anwendung radioaktiver und stabiler Isotope in allen Gebieten der Forschung und Technik, der Industrie, Landwirtschaft und Medizin, Herstellung radioaktiver und stabiler Isotope sowie von markierten Verbindungen, kernelektronische Geräte für die Isotopenanwendung (Dickenmesser, Dichtemesser u. ä.) Detektoren für isotopentechnische Arbeiten und den Strahlenschutz in Isotopenlaboratorien, Meßtechnik und Strahlenschutz, Dosimetrie in Isotopenlaboratorien, kerntechnische Geräte für die Isotopenanwendung (Beton, Bestrahlungsanlagen u. ä.), Strahlenchemie, Radiochemie, ökonomische und perspektivische Fragen, Fragen der Weiterbildung.

1. Aufnahme von Manuskripten

Zur Veröffentlichung angenommen werden nur druckreife, noch nicht anderweitig publizierte Beiträge in deutscher, englischer oder russischer Sprache zum Themenbereich der Zeitschrift und zwar

- Originalarbeiten
- Kurze Mitteilungen
- Aus der Praxis
- Zusammenfassende Berichte
- Tagungsberichte
- Informationen.

Sind der Beitrag oder Teile daraus bereits veröffentlicht oder ist eine anderweitige Veröffentlichung vorgesehen, so muß dies der Redaktion mit genauer Angabe der Zweitveröffentlichung mitgeteilt werden. Mit der Annahme des Manuskriptes behält sich der Verlag das Recht der Vervielfältigung, Verbreitung und Übersetzung innerhalb der gesetzlichen Schutzfrist vor. Auszugsweiser Nachdruck oder photomechanische u. a. Wiedergabe sind an eine schriftliche Genehmigung des Verlages gebunden. Manuskripte, Zeichnungen und Photos von veröffentlichten Arbeiten werden nur auf den besonderen und bis zum Zeitpunkt der Veröffentlichung mitgeteilten Wunsch des Autors zurückgesandt.

Bei der unverlangten Einreichung von Manuskripten wird empfohlen, eine Befürwortung eines der Herausgeber oder Redaktionsausschußmitglieder beizufügen.

Manuskripte sind einzureichen bei der Redaktion, 1086 Berlin Leipziger Str. 3–4.

2. Form der Manuskripte

Die Autoren werden gebeten:

2 Manuskripte, eines davon in Original-Maschinenschrift, einseitig und mit 2zeiligem Abstand einzureichen; gut leserliche Photokopien sind zulässig; auf dem 1. Blatt (unter Weglassung der Bezeichnung)

- den Titel der Arbeit
- Vorname-Initialen und Namen sämtlicher Autoren
- gegebenenfalls die Institution, in der sie tätig sind
- eine kurze Zusammenfassung der Arbeit von höchstens 15 Zeilen
- Keywords in alphabetischer Reihenfolge und
- als Fußnote die Anschrift der Institution anzugeben;

wenn möglich, Zusammenfassung, Titel und Autoren *englisch* (bei deutschem oder russischen Manuskript) bzw. *deutsch* (bei englischem Manuskript) 2fach beizufügen;

Literaturhinweise (fortlaufend numeriert), Fußnoten, Tabellen, Abbildungen (in 2–3facher Veröffentlichungsgröße), Abbildungsunterschriften (getrennt von den Abbildungen und Zeichenerklärungen, Quellenangaben, falls erforderlich, und gegebenenfalls mit einem Vermerk für besondere Wünsche bezüglich der Abbildungsgröße), private und Dienstanschrift der Verfasser mit akademischen Titeln und ausgeschriebenen Vornamen auf besonderen Blättern ans Ende des Manuskriptes zu stellen;

Wünsche für den Kolumnentitel (Seitenüberschrift) bekanntzugeben;

Absätze durch Einzüge von mindestens 3 Anschlägen kenntlich zu machen; die Maßeinheiten des SI-Systems („Systeme International") zu verwenden; Korrekturen im Manuskript unmittelbar an der zu korrigierenden Stelle anzubringen;

Formeln deutlich lesbar und möglichst groß einzutragen;

Farbabbildungen und nachträgliche Änderungen oder Ergänzungen in Abbildungen zu vermeiden;

hervorzuhebende Textstellen zu unterstreichen, und zwar halbfetten Satz mit einer geraden schwarzen Linie, kursiven Satz (Schrägschrift) mit einer blauen Wellenlinie (alle Eigennamen). Sperrungen gestrichelt;

besondere Buchstaben durch farbige Unterstreichung zu kennzeichnen, und zwar Formelzeichen in

- Griechisch-Kursiv (α, β, ...) rot,
- Kursiv (*A*, *B*, ...) blau
- halbfetter Kursiv (**A**, **B**, ...) 2 Linien blau (Vektoren),
- halbfetter Grotesk (**A**, **B**, ...) 2 Linien violett (Tensoren);

bei Literaturangaben auf Vollständigkeit zu achten. In nachstehender Reihenfolge müssen angegeben sein:

a) bei *Zeitschriften*: Vorname-Initialen und Namen sämtlicher Autoren, Titel der Zeitschrift (nach „Periodica Chimica" oder möglichst wenig gekürzt), Bandnummer (unterstrichen), Jahreszahl (in Klammern), Heftnummer (nur bei Zeitschriften, die jedes Heft mit S. 1 beginnen) sowie die Seitenzahl, auf der die zitierte Arbeit beginnt.

b) bei *Büchern*: Initialen der Vornamen und Name des Autors (Herausgeber werden durch den Zusatz „ed." gekennzeichnet), voll ausgeschriebener Titel des Buches in der Originalsprache, evtl. Bandnummer, voll ausgeschriebener Name des Verlages, Nummer der Auflage (außer der ersten), Erscheinungsort(e), Erscheinungsjahr (hier nicht in Klammern setzen!) und evtl. Kapitelnummer, Seitenzahl u. ä.

c) bei *Sammelbänden, Tagungsberichten*: Es soll zusätzlich der Titel des zitierten Beitrags angegeben werden. Die Seitenzahl muß unbedingt angegeben werden. Aus dem Zitat muß hervorgehen, ob es sich um einen Tagungssammelband oder um einen noch nicht veröffentlichten Vortrag handelt. Zwischen Tagungsort und Erscheinungsjahr des entsprechenden Tagungsberichts ist zu unterscheiden.

3. Korrekturen

Der Autor erhält zwei Umbruchsabzüge, von denen einer möglichst schnell korrigiert an die Redaktion zurückgereicht werden soll. Auf Wunsch des Autors kann die Korrektur auch von der Redaktion gelesen werden.

Durch nachträgliche Manuskriptänderungen verursachte Korrekturkosten gehen, sofern sie 10% der Satzkosten übersteigen, zu Lasten des Autors.

Der Autor erhält 50 Sonderdrucke kostenlos. Weitere Sonderdrucke können bis zur Rückgabe der Korrekturabzüge auf Kosten des Autors bei der Redaktion bestellt werden.

Bezugsmöglichkeiten

Bestellungen sind zu richten

— **in der DDR** an den Postzeitungsvertrieb unter Angabe der Kundennummer des Bestellenden oder an den AKADEMIE-VERLAG BERLIN, DDR-1086 Berlin, Leipziger Straße 3–4, PF-Nr. 1233;
— **im sozialistischen Ausland** an eine Buchhandlung für fremdsprachige Literatur oder an den zuständigen Postzeitungsvertrieb;
— **in der BRD und Berlin (West)** an eine Buchhandlung oder an die Auslieferungsstelle KUNST UND WISSEN, Erich Bieber OHG, Wilhelmstraße 4–6, D-7000 Stuttgart 1;
— **in den übrigen westeuropäischen Ländern** an eine Buchhandlung oder an die Auslieferungsstelle KUNST UND WISSEN, Erich Bieber GmbH, General Wille-Str. 4 CH-8002 Zürich;
— **im übrigen Ausland** an den Internationalen Buch- und Zeitschriftenhandel; den Buchexport Volkseigener Außenhandelsbetrieb der Deutschen Demokratischen Republik, DDR-7010 Leipzig, Postfach 160; oder an den AKADEMIE-VERLAG BERLIN, DDR-1086 Berlin. Leipziger Straße 3–4.

Zeitschrift „Isotopenpraxis"

Herausgeber: Zentralinstitut für Isotopen- und Strahlenforschung der Akademie der Wissenschaften der DDR.
Verlag: Akademie-Verlag Berlin, DDR-1086 Berlin, Leipziger Straße 3–4, PF-Nr. 1233; Fernruf: 223621 oder 2236229; Telex-Nr.: 114420;
Bank: Staatsbank der DDR, Berlin, Kto.-Nr.: 6836-26-20712.
Chefredakteur: Prof. Dr. H. Hübner, Leipzig.
Anschrift der Redaktion der Zeitschrift Isotopenpraxis: Prof. Hübner, DDR-7010 Leipzig, Karl-Marx-Platz, Hauptpostamt, PSF 370; Fernruf: 281018.
Veröffentlicht unter der Lizenznummer 1527 des Presseamtes beim Vorsitzenden des Ministerrates der Deutschen Demokratischen Republik.
Gesamtherstellung: VEB Druckerei „Thomas Müntzer", DDR-5820 Bad Langensalza.
Erscheinungsweise: Die Zeitschrift „Isotopenpraxis" erscheint monatlich. Die 12 Hefte einer Jahrganges bilden einen Band. Bezugspreis je Band 240,— DM zuzüglich Versandspesen; Preis je Heft 20,— DM. Der gültige Jahresbezugspreis für die DDR ist der Postzeitungsliste zu entnehmen.
Bestellnummer dieses Heftes: 1079/22/2.

AN (EDV) 12831
00900

Isotopenpraxis 22 (1986) 2, S. 41 bis 45

Prediction of Removal Efficiency of Rectangular Settling Tank
Part I. Application of Radiotracer Method in Tank Testing and Laboratory-Scale Verification of the Mathematical Model of Sediment Transport

St. Szpilowski[1]), *J. St. Michalik*[2]), *A. Owczarczyk*[1])

([1]) Institute of Nuclear Chemistry and Technology, Department of Nuclear Methods for Process Engineering, Warszawa; [2]) Institute of Atomic Energy Otwock-Świerk)

Among many factors affecting processes of the suspension removal in settling tanks an important role play turbulent mixing and resuspension of sediment deposited on the bottom of tank. For the purposes of removal process analysis the mathematical model of suspended sediment transport has been established. The model has been used for prediction of removal efficiency of laboratory-scale rectangular settling tank. For evaluation of parameters characterizing flow patterns and sediment falling velocity the radiotracer method has been used. The application of the model has been verified in laboratory conditions with satisfactory results. That gives base for planned study of industrial full scale rectangular settling tanks.

Unter den Faktoren, die den Prozeß der Suspensionsabtrennung in Abscheidetanks beeinflussen, spielen die turbulente Mischung und die Resuspension des abgelagerten Sediments eine wichtige Rolle. Zu diesem Zweck wurde ein mathematisches Transport-Modell für das suspendierte Sediment entwickelt und für Voraussagen der Abtrennungseffektivität im Labormaßstab verwendet. Zur Ermittlung der Parameter, die den Fluß und die Sinkgeschwindigkeit des Sediments charakterisieren, wurde die Radiotracermethode eingesetzt. Die Anwendung des Modells ergab für den Labormaßstab befriedigende Ergebnisse, und sie bilden die Grundlage für einen industriellen rechteckigen Abscheidungstank.

Keywords

mathematical models; radioisotopes; scale dimension; scale models; sedimentation; suspensions; tracer techniques

1. Introduction

Removal of suspended matter from sewage is one of stages of sewage purification and occurs usually prior to chemical or biological treatment. Some systems use this also as a final stage based on separation of purifying agent e.g. activated sludge. Removal of solids becomes important stage of a process of sewage purification. Sometimes this is crucial for effective elimination of remaining pollutants. The purpose of this work is to find a method of evaluating the suspended matter removal efficiency in settling tank under various operating conditions and prediction of operating efficiency for design purposes.

The work also includes investigation of suspended sediment transport processes which require assessment of both liquid flow and settling process parameters. Both processes have been investigated by means of radiotracer method.

Tracer methods provide valuable information in industrial processes and equipment research. Commonly used method consists in analysing system response to interfering signal and is particularly useful in single — or multi — phase flow systems analysis. The introduction of radioisotopes as tracers considerably widened the field of application of the methods. Detection of radioactive tracer is not interfered by such factors as chemical composition variations of flowing medium, presence of foreign phases or changes of colour. These factors make measurements by means of conventional tracers (e.g. dyes or salts) considerably difficult or even impracticable. Another advantage of radioactive tracers is the possibility of their high initial activity in rather small quantity. Introduction of small mass of a tracer does not provide any disturbance to production process being investigated. Insensibility of radioactive tracers detection to variations of physical properties of surrounding environment is of particularly great value for water and sewage research. The application of radioisotopes provides quick and undisturbed determination of dynamic characteristics of tested installations and process parameteres.

2. Selection of radioactive tracer

Tracer used in industrial processes investigation should meet several requirements in respect to its physical and chemical properties and free-space pattern.

Tracer introduced to the system under test should be involved selectively in those processes which are the matter of interest. It is essential that the tracer should not pass to the phase other than labelled. Moreover, the half-life period of a tracer should be from several hours to several tens of hours. This is sufficiently long time to deliver it to the site and to perform experiments without prolonged contamination of examined medium.

A ^{82}Br [1] isotope in form of KBr and ^{99m}Tc in form of NH_4TcO_4 have been eventually selected as water tracers. ^{82}Br is obtained by activation of potassium bromide or ammonium bromide in a nuclear reactor. This tracer was used in investigations of industrial settling tanks.

Addresses: [1]) Dorodna 16, 03-195 Warszawa; [2]) 05-400 Otwock-Świerk-Poland

^{99m}Tc is received from the $^{99}Mo/^{99m}Tc$ generator and used in laboratory tests. This nuclide emits low energy γ radiation and its industrial application is limited by high manufacturing cost of such generator which can provide tracer of sufficiently high activity.

^{113m}In in form of $InCl_3$ was used as a tracer of sediments. This tracer is received from the $^{113}Sn/^{113m}In$ generator. Tracing is made through sorption of ^{113m}In on sediment surface.

3. Theoretical backgrounds for sediment transport process analysis

Sewage purification process occuring in settling tanks consists in phenomena related to nature of liquid flow through installation and sediment settling.

The efficiency of purification process depends in this case both on type of solids being settled and design parameters of settling tank as well as its sewage charge.

Required removal efficiency can be achieved by such tank design which can provide sufficient retention time of waste water in installation. In this respect turbulent flow is most undesirable since it affects sedimentation and in extreme cases stops it completely. Turbulence results in entrainment of sediment particles already settled on the bottom of tank. The resuspension process can be described by the following coefficient:

$$K = \frac{n'}{n} \qquad (1)$$

where: n' — number of sediment particles entrained into flow zone per unit of time
n — number of sediment particles reaching the tank bottom per unit of time

The above assumptions lead to the dispersion model of suspended matter transport [2].

The model equation for settling tank having length L and depth H in steady state conditions is as follows:

$$D_x \frac{\partial^2 c}{\partial x^2} + D_z \frac{\partial^2 c}{\partial z^2} = u \frac{\partial c}{\partial x} - v \frac{\partial c}{\partial z} \qquad (2)$$

with boundary conditions:

$$x = 0 \qquad c = c_0 \qquad (3)$$

$$x = L \qquad \frac{\partial c}{\partial x} = 0 \qquad (4)$$

$$z = 0 \qquad D_z \frac{\partial c}{\partial z} + Kvc = 0 \qquad (5)$$

$$z = H \qquad D_z \frac{\partial c}{\partial z} + vc = 0 \qquad (6)$$

where: x, z — length and depth co-ordinates
c — sediment concentration
u — mean linear flow velocity
v — mean linear settling velocity

Solution of equation (2) describes distribution of solids concentration in a settling tank for constant parameters of flow (D_x, D_z, u) and sedimentation (v, K).

Since measurement of vertical dispersion coefficient D_z is extremely difficult and highly inaccurate, it seems resonable to introduce certain simplifications consisting in concentration averaging along the depth.

Thus the equation (2) can be reduced to one-dimensional form by substitution of the formula:

$$\bar{c} = \frac{1}{H} \int_0^H c(x, z)\, dz \qquad (7)$$

If we assume that sediment concentration in surface and bottom layers is nearly equal to the average sediment concentration in settling tank cross-section area, then the following equation of transport is derived:

$$D_x \frac{d^2 \bar{c}}{dx^2} - u \frac{d\bar{c}}{dx} - \frac{v}{H}(1 - K)\, c = 0 \qquad (8)$$

with boundary conditions:

$$x = 0, \qquad \bar{c} = c_0 \qquad (9)$$

$$x = L, \qquad \frac{d\bar{c}}{dx} = 0 \qquad (10)$$

The equation (8) is a second order ordinary differential equation which can be easily solved.

Thus the removal efficiency (R) is determined from the formula:

$$R = 1 - \frac{\varphi - \psi}{\varphi \exp(-\psi) + \psi \exp(-\varphi)} \qquad (11)$$

where:

$$\varphi = \frac{P_e}{2}\left\{1 + \left[1 + \frac{4}{P_e}(1 - K)\frac{v}{u}\frac{L}{H}\right]^{1/2}\right\} \qquad (12)$$

$$\psi = \frac{P_e}{2}\left\{1 - \left[1 + \frac{4}{P_e}(1 - K)\frac{v}{u}\frac{L}{H}\right]^{1/2}\right\} \qquad (13)$$

$$P_e = \frac{uL}{D_x} \qquad (14)$$

The equation (11) represents relationship between removal efficiency, flow and settling parameters.

To use the equation (11) for predicting removal efficiency on the grounds of known settling tank dimensions, sewage flow rate and mean settling velocity it is necessary to know interrelations between transport and settling parameters. These interrelations should be taken from experiment.

4. The method of measurement of dispersion coefficient

The dispersion coefficient is determined by means of analysis of recorded tracer concentration as a function of time at inlet and outlet of settling tank. Residence time distribution curves obtained in this way are then taken as input signal and response curves.

Recorded response function is given by convolution integral:

$$y(t) = \int_0^T x(t - \tau)\, f(\tau)\, d\tau \qquad (15)$$

where:

$x(t)$ — input signal
$y(t)$ — response function
$f(t)$ — impulse response function

The convolution equation is solved by means of *Fourier* transform [3—5]. The algorithm of fast *Fourier* transform [5—7] has been applied for calculation of impulse-response function.

To determine the actual value of dispersion coefficient on the base of analysis of measured input signal and response curves the 4-parameter-dispersion model was used. Known model transfer function has been used [8, 9]

$$F(s, p) = F(s, \tau, Pe, k, \alpha) =$$

$$= \exp\left\{\frac{Pe}{2}\left[1 - 1 + \frac{4s\tau}{Pe}\left[1 + \frac{\alpha k}{\alpha s + k}\right]\right)^{1/2}\right]\right\} \quad (16)$$

where:

$$\tau = \frac{L}{u} \quad (17)$$

$$Pe = \frac{ul}{D_x} \quad (18)$$

$$s = 2\pi i v \quad (19)$$

α – fraction of dead space
k – mass transfer coefficient between dead space and flow region
L – distance between measuring points
u – mean linear flow velocity

Values τ, Pe, α, k are model parameters. Function $F(s, p)$ from equation (16) is a impulse response transfer function.

The values of model parameters have been obtained by means of least square fittling procedure [8] in frequency domain.

The measurements were carried out in a laboratory rectangular settling tank. The tank was made from perspex glass and supplied with water from an intermediate tank by means of automatic control system with flow rate measuring means. In order to equalize flow velocity distribution perforated baffles were installed in the inlet part of the tank. Under the bottom of the tank, near its inlet and outlet, two collimated scintillation probes were installed. They were connected with counting and recording system.

Measured input signal and response curves were recorded in the form ready for the data treatment processor. The tracer (^{99m}Tc) was injected to the supply pipeline. The tracer activity was 0·05 GBq (1·35 mCi). To determine the relation between the dispersion coefficient and hydraulic parameters of the tank a series of isotope measurements concerning residence time distribution curves for various flow rates through the tank (Q) and various depths (H) was performed. Fig. 1 shows the scheme of measuring procedure. Recorded input signal- and response standard curves were treated according to the procedure described earlier. The results obtained permit to establish relation

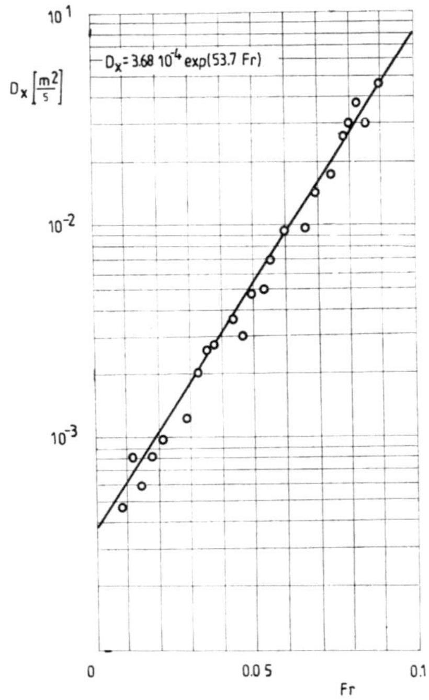

Fig. 2. Values of dispersion coefficient and regression line $D_x = 3{,}68 \times 10^{-4} \exp(53{,}7\ Fr)$

between the longitudinal dispersion coefficient and *Froude* number in form of regression line.

$$D_x = 3{\cdot}68 \cdot 10^{-4} \exp(53{\cdot}7\ Fr) \quad (20)$$

where:

$$Fr = \frac{u}{(gH)^{1/2}} \quad (21)$$

g – acceleration of gravity

Fig. 2 shows measured D_x values and a regression line (20).

5. Determination of settling velocity

Suspended calcium carbonate was used as a sediment for this study. The measurements of settling velocity were carried out on vertical column filled up to the height of 1·5 m with solids/water mixture with concentration of 2 g/l. The 110 mm diameter column was made of perspex glass. The sediment was labelled by adding the ^{113m}In

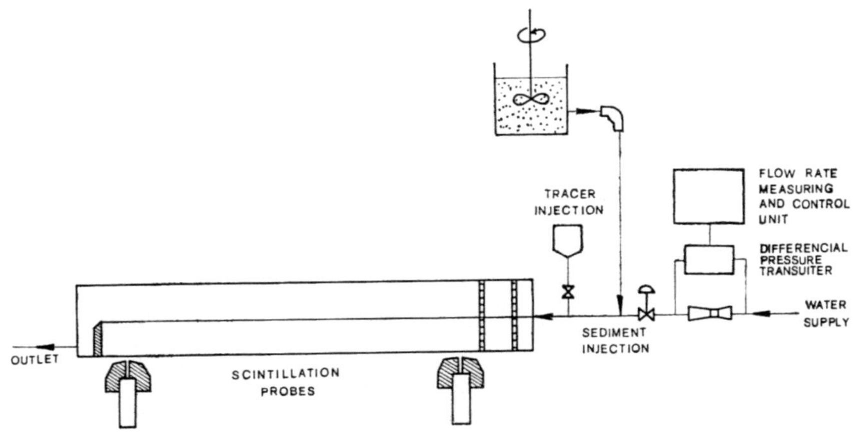

Fig. 1. Scheme of laboratory installation

tracer of activity of 0·1 GBq (2·7 mCi). The column content was then mixed by means of compressed air.

Two collimated scintillation probes were fitted to the outer column wall at levels of 1/3 and 2/3 of column height. They were connected with measuring and recording system, the same as described in p. 4.

The measurement of settling time distribution curves was initiated at the instant cut-off of compressed air supply to the column.

Fig. 3 shows the curves $R = f(t)$ obtained from measurements. The determined value of settling linear velocity amounted to $v = 2{\cdot}3 \cdot 10^{-4}$ m/s.

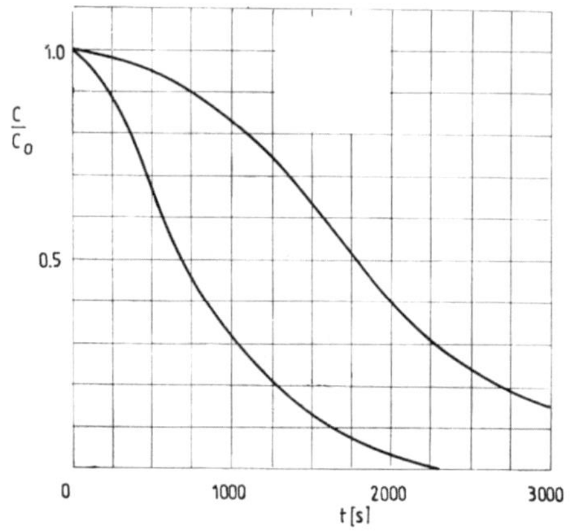

Fig. 3. Measured curves of removal time distribution
$h = 0{\cdot}5$ m; $t = 2174$ s; $v = 2{\cdot}3 \cdot 10^{-4}$ ms^{-1}

6. Determination of resuspension coefficient

The direct measurement of the coefficient K is not possible. This value was determined from the equation (11). In order to attain this, a series of isotope measurements on model tank were carried out included determination of removal efficiency with simultaneous measurement of flow parameters (u, D_x). The tank was supplied with water and calcium carbonate suspension labelled with ^{113m}In. Initial sludge concentration in the tank c_0 was about 2 g/l. The removal efficiency was determined by sampling at tank inlet and outlet sections with measurement of sample activity. At the same conditions an isotop measurement of flow parameters according to procedure described in p. 4 was performed. On the grounds of the data obtained the relation between resuspension factor (K) and longitudinal dispersion coefficient in form of regression line

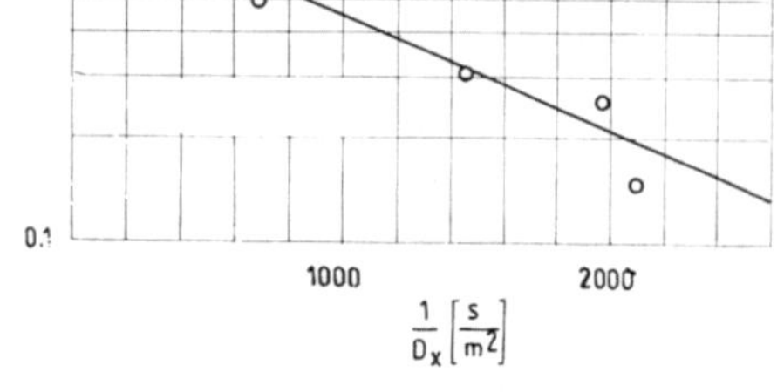

Fig. 4. Values of resuspension coefficient and regression line $K = 9{\cdot}56 \cdot 10^{-4} \exp\left(-\frac{7{,}52}{D_x} \cdot 10^{-4}\right)$

has been determined:

$$K = 0{\cdot}956 \exp\left(-\frac{7{\cdot}52 \cdot 10^{-4}}{D_x}\right) \tag{22}$$

Fig. 4 shows measured values K together with the regresion line (22).

7. Verification of the model

The equations (11), (20) and (22) describe a sediment transport model. They can be used to predict removal efficiency related to basis dimensions of a tank (L, H) and known linear flow velocity (u) and settling parameters (v, K). Fig. 5 shows removal efficiency as a function of tank depth for various flow rates, determined from model equations.

Fig. 6 shows the curves $R = f(H)$ for tanks of various length. The effect of resuspension factor on calculated

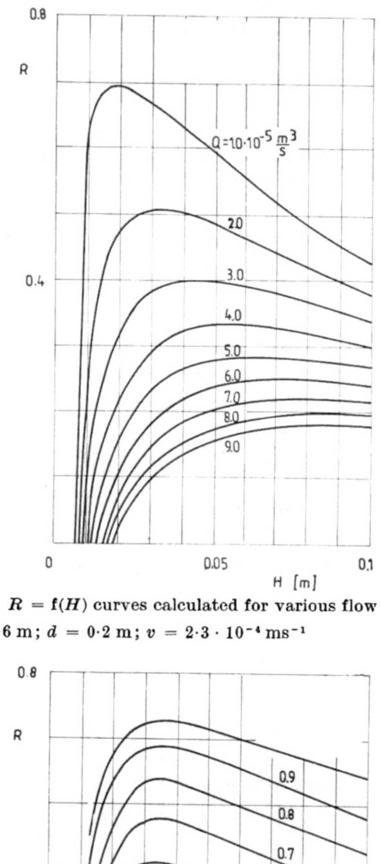

Fig. 5. $R = f(H)$ curves calculated for various flow rates
$L = 0{\cdot}6$ m; $d = 0{\cdot}2$ m; $v = 2{\cdot}3 \cdot 10^{-4}$ ms^{-1}

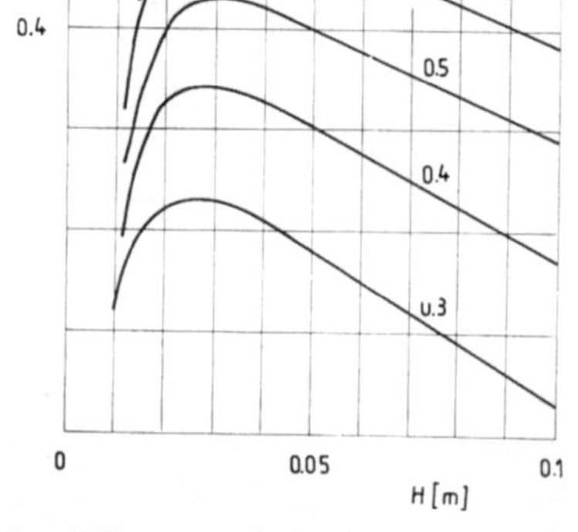

Fig. 6. $R = f(H)$ curves calculated for various length of the tank
$Q = 2 \cdot 10^{-5}$ m$^3 \cdot$ s^{-1}; $d = 0{\cdot}2$ m; $v = 2{\cdot}3 \cdot 10^{-4}$ ms^{-1}
Addition: At the top curve read $L = 1 \cdot 0$ m

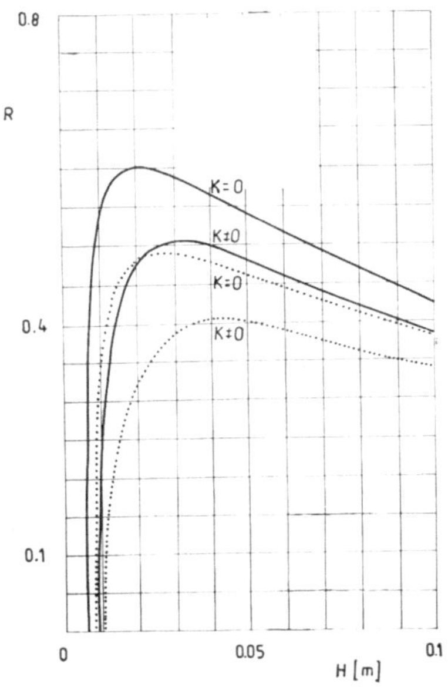

Fig. 7. $R = f(H)$ curves calculated for $K \neq 0$ and respective curves for $K = 0$
$L = 0{\cdot}6$ m; $d = 0{\cdot}2$ m; $v = 2{\cdot}3 \cdot 10^{-4}$ ms^{-1}; ——— $Q = 2{\cdot}0 \cdot 10^{-5}$ m^3 × × s^{-1}; ······ $Q = 3{\cdot}0 \cdot 10^{-5}$ m^3 s^{-1}

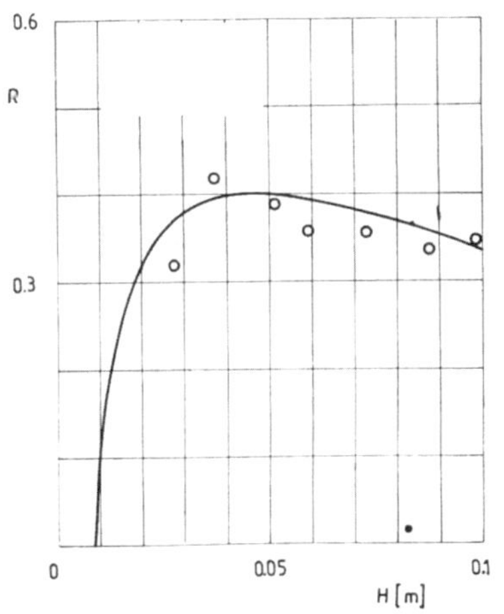

Fig. 8. Measured values of removal efficiency and resective predicted curve $R = f(H)$
$L = 0{\cdot}6$ m; $d = 0{\cdot}2$ m; $v = 2{\cdot}3 \cdot 10^{-4}$ ms^{-1}; $Q = 3{\cdot}0 \cdot 10^{-5}$ m^3 s^{-1}

value of removal efficiency is shown is Fig. 7. Two curves $R = f(H)$ calculated from the equations (11), (20) and (22) and respective curves for $K = 0$ are shown.

To compare calculated values of removal efficiency with experimental results two sets of R value determinations for model settling tank according to description in p. 6 were performed. The results obtained are presented in Fig. 8 and 9.

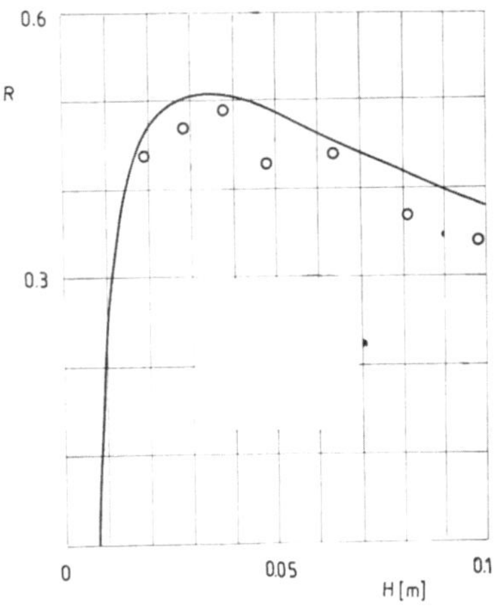

Fig. 9. Measured values of removal efficiency and resective predicted curve $R = f(H)$
$L = 0{\cdot}6$ m; $d = 0{\cdot}2$ m; $v = 2{\cdot}3 \cdot 10^{-4}$ ms^{-1}; $Q = 2{\cdot}0 \cdot 10^{-5}$ ms^{-1}

8. Conclusions

The experimental results show satisfactory conformity with values calculated from equations (11), (20), (22).

Measured removal efficiency values are usually less than calculated, what can be attributed to the fact that not all phenomena associated with waste purification process occurring in settling tank have been included in applied mathematical model of sediment transport and fluid flow.

The method presented requires verification in industrial practice.

Laboratory work was carried out in conditions of constant water flow through sediment tank and equalization of flow linear velocity in the cross section of the tank.

Such steady state operation conditions of commercial settling tanks are impracticable and therefore considerable discrepancies between measured and calculated values of removal efficiency have to be expected.

Final suitability of the proposed method will result from measurement error analysis for commercial applications.

Received May 9, 1985

References

[1] *W. Zabłotny, R. Winnicki*, Nukleonika **16** (1971) 389
[2] *St. Szpilowski*, "Application of the Axially Disperced Plug Flow Model for Description of Sewage Treatment Process in Rectangular Settling Tank", Institute of Nuclear Research (Manuscript), 1971
[3] *I. N. Sneddon*, "The Use of Intergral Transforms". McGraw-Hill Book Comp. 1972
[4] *W. T. Cochran, J. J. Downing, D. L. Davin, H. D. Helms, R. A. Kaenel, W. W. Lang, D. E. Nelson*, Burst Measurements in the Frequency Domain, Proc. IEEE, Vol. **54** (1966) p. 830
[5] *W. T. Cochran, J. W. Cooley, D. L. Favin, H. D. Helms, R. A. Kaenel, W. W. Lang, G. C. Maling, D. E. Nelson, C. M. Rader, P. D. Welch*, "What is the Fast Fourier Transform". Proc. IEEE, Vol **Au-15** (1967) No. 2, p. 45
[6] *J. W. Cooley, J. W. Tukey*, Math. Comput. **19** (1965) p. 297
[7] *J. W. Cooley., A. W. Lewis, P. D. Welch*, "Application of the Fast Fourier Transform to Computation of Fourier Integrals, Fourier Series and Convolution Integrals". Trans IEEE, Vol. **Au-15** (1967) No. 2, p. 79
[8] *M. L. Michelsen*, Chem. Eng. J. **4** (1972) p. 171
[9] *J. Villemaux, W. P. Van Swaail*, Chem. Eng. Sci. **24** (1969) p. 1097

Isotopenpraxis 22 (1986) 2, S. 46 bis 49

Prediction of Removal Efficiency of Rectangular Settling Tank Part II. Investigation of Commercial Scale Settlers

St. Szpilowski[1]), *J. St. Michalik*[2]), *A. Owczarczyk*[1])

([1]) Institute of Nuclear Chemistry and Technology, Department of Nuclear Methods for Process Engineering, Warszawa; [2]) Institute of Atomic Energy Otwock-Swierk)

The work on the verification of the method of the evaluation of removal efficiency in rectangular settling tanks, is presented. The method has been checked earlier in laboratory scale [1]. The experimental work was carried out on a preliminary settling tank and on a secondary settler of activated sludge. The method was found suitable for sediments which do not undergo flocculation.

Es wird eine vorgeschlagene Methode zur Bestimmung der Abscheidungseffektivität in rechteckigen Abscheidern geprüft. Die Methode wurde vorher im Labormaßstab getestet [1], und die Experimente wurden an einem vorgeschalteten Abscheidungstank gekoppelt mit einem Belebtschlammbecken ausgeführt. Die Methode erwies sich als geeignet für Sedimente, die keiner Auflockerung unterliegen.

Keywords

flocculation; indium 113m; radioisotopes; sedimentation; slurries; suspensions; tracer techniques

1. Introduction

The process of elimination of suspended matter from wastes in settling tanks depends on a number of factors. The most important of which are the tank load and the type of sediment being removed, as well as the characteristics of the fluid flow through the installation.

The Institute of Nuclear Chemistry and Technology is involved in work on the method of predicting the removal efficiency of rectangular settling tanks. The work is based on the assumption that longitudinal mixing is a factor disturbing the process of sedimentation together with resuspension of particles deposited on the bottom of the tank.

The relations obtained have been tested with positive result for predicting the removal efficiency in laboratory scale [1]. In order to determine the suitability of the method for the analysis of the operation of settling tanks in sewage treatment plants, it is necessary to check the elaborated relations in practical applications.

2. Experimental

The method presented can be used for sediments which do not undergo flocculation. If the flocculation affects the settling process it may be expected that the values obtained from model equations [1] will vary from the observed values of removal efficiency.

The experimental work carried out on a preliminary and secondary settling tanks represent two extreme types of installations. The first one is a sand trap and teap, and is used for preliminary separation of coarse sludge, while the latter is a secondary settling tank and is used for separation of the activated slugde from treated sewage. Since the flocculation plays an important role in the settling process of activated sludge, the examination of this type of settlers facilitates the checking of discrepancy between predicted and measured values of removal efficiency under conditions different from these initially included within the range of applicability of the method.

2.1. *Testing of settling processes*

The measurements of the linear velocity of settling of the sediments (v) have been carried out in a vertical column filled with a mixture of sewage and sludge taken from the channel supplying the sewage to the tank. The measurement was performed by means of radiotracer method, by labelling the sludge with ^{113m}In isotope. The settling time distribution curves were recorded by means of two collimated scintillation probes placed at fixed points of external walls of the column.

In order to determine the relation between the values of the resuspension coefficient (K) and the longitudinal dispersion coefficient (D_x), a series of determinations of the removal efficiency in a laboratory installation was performed. The installation was supplied with mixture of sewage and sludge. Simultaneously, the value of the longitudinal dispersion coefficient was measured using radioactive tracer [1].

Fig. 1 and 2 present the determined regression lines $K = f(D_x)$ for the examined sediments.

2.2. *Testing of industrial installations*

The measurement of the values of longitudinal dispersion coefficient consisted in the injection of the radioactive tracer into the inlet channel of the tank, and recording of signal and response curves at the inlet and outlet of the tank. The ^{82}Br in the form of KBr was used as a tracer.

The signal and response curves are standardized retention time distribution (RTD) functions at the inlet and outlet of the installation under test. The measurement of RTD-curves requires the recording of average tracer concentration in the cross-section area of the inlet and outlet stream as a function of time. If a radioactive tracer is used, the

Addresses: [1]) Dorodna 16, 03-195 Warszawa; [2]) 05-400 Otwock Świerk-Poland

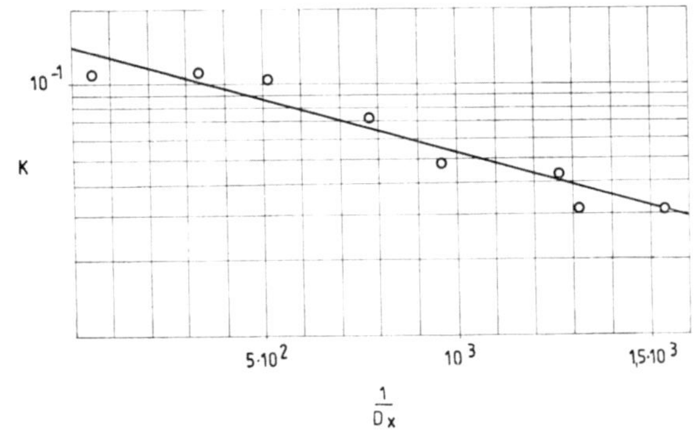

Fig. 1. The primary settling tank
values of the resuspension coefficient K and corresponding regression line $K = 1 \cdot 44 \cdot 10^{-1} \exp\left(-\frac{1}{D_x} \cdot 10^{-3}\right)$

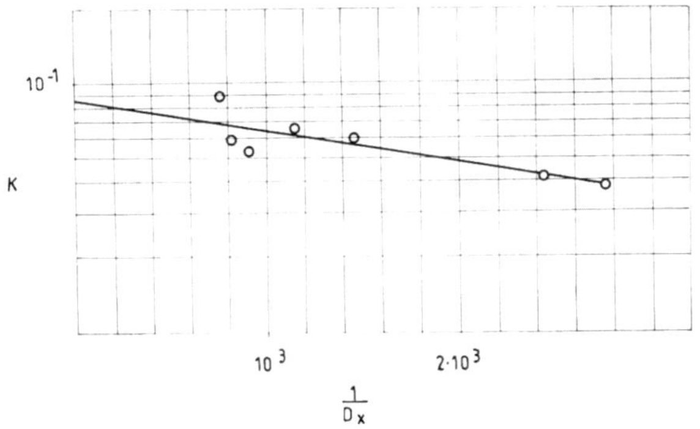

Fig. 2. The secondary settling tank
values of the resuspension coefficient K and corresponding regression line $K = 8{\cdot}54 \cdot 10^{-2} \exp\left(-\frac{2{\cdot}85}{D_x} \cdot 10^{-4}\right)$

changes of its activity at suitable points of measurement are recorded. Because of the dimensions of commercial installations, it becomes necessary to record these curves simultaneously at a number of points in the cross-section area at given measurement section.

In the conducted work, apparatus facilitating simultaneous numerical recording of counts from a maximum number of fifteen points was used.

The apparatus consists of a set of fifteen scintillating probes NaJ/Tl together with H.V. power supply units and single-channel analyzers, five scalers equipped with internal and buffer store, a control block and paper tape puncher. Each scaler records the counts in parallel from three probes at given time interval. Additionally, the measuring set is equipped with a system of integrators and recorders facilitating direct observation of the changes of tracer activity at the separate test points.

The apparatus counts the pulses simultaneously from all probes at the given time interval. The minimum time interval for counting is 1 second. The buffer store facilitates continuous measurement, i.e. without time breaks between the separate periods of counting.

The puncher output has the form ready for data treatment processor.

The sinctillation probes were located at two measuring sections at the inlet and outlet sections of the settling tank. The arrangement of the measuring probes is shown in Fig. 3 and 4.

Fig. 3. The primary settling tank, the layout of measuring points
1 – scintillation probes

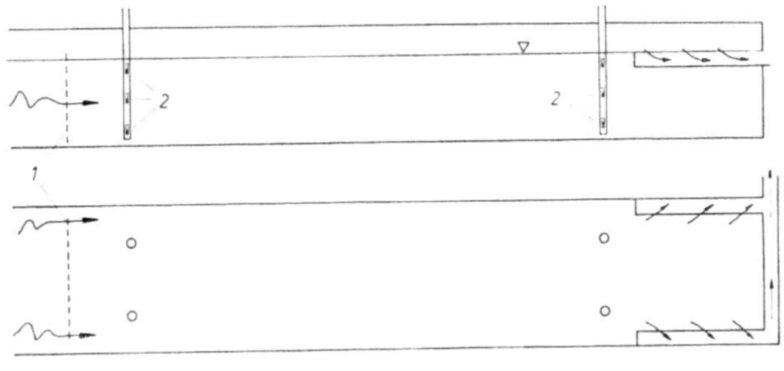

Fig. 4. The secondary settling tank, the layout of measuring points
1 – partition; 2 – scintillation probes

A series of 28 measurements have been performed on the primary settling tanks for various loadings and depth of the tank (the change of the depth was achieved by the adjustment of the heigh of the outlet overflow). The activity of the tracer used for a single measurement was 0·5 GBq (13·5 mCi) ^{82}Br.

On the secondary settling tank two measurements at a constant load and constant level of liquid were performed. The first measurement was made under normal operating conditions. For the time of the second measurement, the partition separating the tank from the remaining part of the treatment plant was removed. The activity of the tracer used for single injection was 2·0 GBq (54 mCi).

For each measurement, the actual removal efficiency was obtained directly by taking samples from the inlet and outlet part of the tank.

The dispersion coefficients have been determined by analysis of the recorded signal and response curves. The mean number of counts in a given measuring profile in a given counting time interval was regarded as proportional to the concentration of the tracer in the cross-section area of the tank. The used procedure of calculations [1] covers the shape of the input signal and the presence of dead spaces as well as the exchange process between the dead space and flow zone.

3. Evaluation of method error

On the determination of the parameters of fluid flow through the settling tank, a possibility of serious errors should be taken into account. This regards in particular the determined values of the longitudinal dispersion coefficient. The adopted mathematical flow model does not cover all the phenomena accompanying the process of fluid flow through the tank. Despite the use of a four-parameter dispersion model, it was expected that the measured values of the dispersion coefficient would be greater than its actual value. In the error analysis, the reference system was adopted as the relation between the values of the dispersion coefficient and *Froude* number. Fig. 5 presents the values of the of longitudinal dispersion coefficient

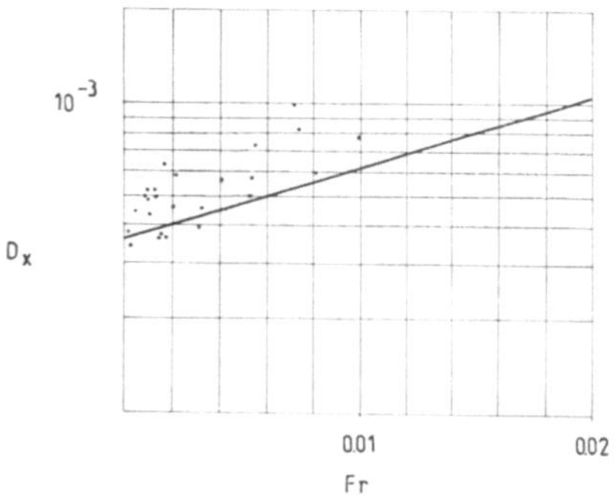

Fig. 5. The primary settling tank
Measured values of the longitudinal dispersion coefficient and regression line $D_x = 3{\cdot}68 \cdot 10^{-4} \exp 53{\cdot}7\, Fr$ determined during laboratory investigations

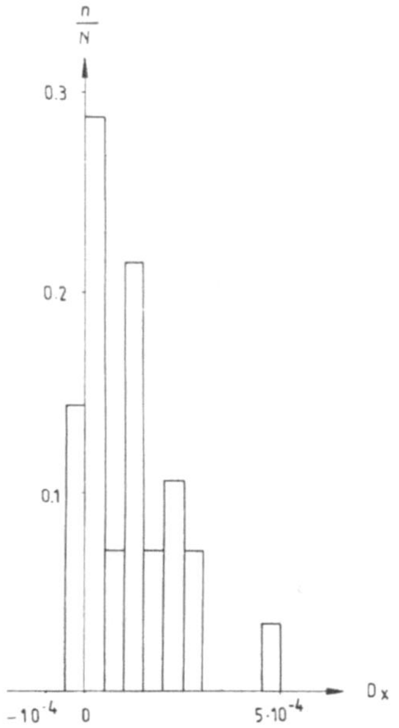

Fig. 6. Ranges of deviations of the measured values of the dispersion coefficient from the regression line $D_x = 3{\cdot}68 \cdot 10^{-4} \exp (53{\cdot}7\, Fr)$ as a function of frequency

measured during investigations of primary settling tank, and the regression line $D_x = 3{\cdot}36 \times 10^{-4} \exp (53{\cdot}7\, Fr)$ determined during laboratory investigations. Almost all measured values of D_x turn out above the regression line. This can be explained by the fact that the adopted model of flow does not cover the stochastic changes of the flow rate of the liquid through the installation and the changes of its density. The changes of the sewage density result in the transient density currents, which affect the shape of the measured curve of the system response by lengthening its "tail". This phenomenon is read by the model as an effect of longitudinal dispersion and liquid exchange between the dead space and the flow zone. As a result, the determined values of the parameters are higher than the actual ones.

Fig. 6 presents the ranges of deviation of the measured values of dispersion coefficients from the regression line $D_x = 3{\cdot}36 \times 10^{-4} \exp (53{\cdot}7\, Fr)$ as a function of frequency of their occurrence. The ratio of the number of measured values of D_x in a given range of deviation to the total number of measurements of the value of D_x has been adopted as a measure of frequency. The value of relative deviation of a highest probability is +7 per cent. The average deviation, under conditions of the investigated values of D_x corresponds to a positive relative deviation of 30–40 per cent. The relative deviations of the D_x values measured in the secondary settler are +14 per cent and +30 per cent.

The evaluation of the R value error determined by means of the suggested mathematical model can be performed by the analysis of the deviations of the predicted values of removal efficiency (R) and actual values (R_c) measured.

Because of the measurement method of the removal efficiency, by direct determination of the concentration of sediment at the inlet and outlet of the settling tank, these values are regarded as accurate ones and adopted as a reference system in the evaluation of the error.

Fig. 7 presents the distribution of the deviations as a function of frequency for values determined for primary settling tank. Because of the similar occurence probability of positive and negative deviation, the distribution of their

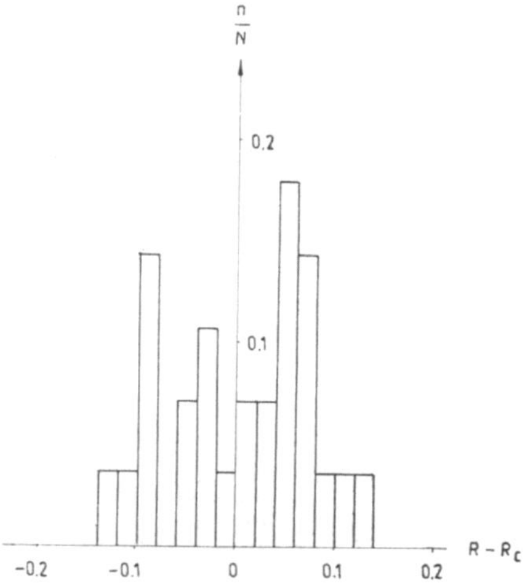

Fig. 7. Ranges of deviations of the measured values of the removal efficiency as a function of frrquency

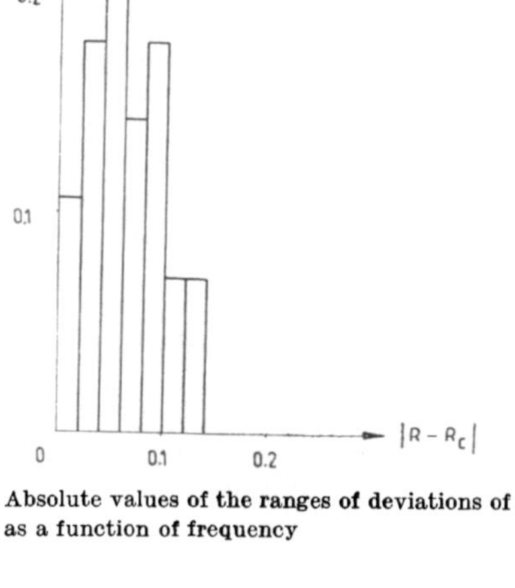

Fig. 8. Absolute values of the ranges of deviations of removal efficiency as a function of frequency

absolute values can be analysed. This distribution is presented in Fig. 8. The average deviation of the R values determined by the use of the model of sediment transport from the measured values is $\pm 0{\cdot}06$ which in the range of R values higher than 0·6 corresponds to a relative deviation of ± 10 per cent. For the results obtained for secondary settling tank, similar deviations amount to 16 per cent and 23 per cent.

4. Summary and conclusions

The presented method of predicting the removal efficiency in rectangular settling tanks is based on a dispersion model of sediment transport [1] and relations between the parameters of the flow and settling processes. The main source of errors is the imperfection of the adopted mathematical models, which do not cover the transient processes taking place in a flow. Despite this, the conformity of the results obtained for the preliminary settling tank can be regarded as satisfactory. The results obtained in the case of activated sludge show an approx. double divergence. It seems that in this case flocculation process plays an important role. The authors see the necessity of continuation of the work for further improvement of the mathematical models used to take into account the processes which occur during sedimentation of various types of sediments.

At present, the method can be used in design work concerning the assumed structure of rectangular settling tank and it permits the determination of the optimum length and depth of the tank on the base of knowledge of the expected load of the installation, the data concerning the type of sediment being removed and the basic parameters of settling process. The application of the method for these purposes requires, each time, the examination of the settling process in laboratory installations.

Received May 9, 1985

References

[1] *St. Szpilowski, J. St. Michalik, A. Owczarczyk* Isotopenpraxis 22 (1986) 41

Isotopenpraxis 22 (1986) 2, S. 49 bis 53

Identification of the Molten Glass Flow Process in Tank Furnaces

K. G. Kruszewski, J. St. Michalik

(Institute of Atomic Energy, Otwock-Świerk)[1]

Classical mathematical model of the flow for a valuation of dynamics of the molten glass flow within the tank furnaces and in compliance with the complex flow model has been analyzed. Investigations of the model of the tank furnace by impulse-response method using ^{140}La radionuclide has been discussed. Obtained technological and dynamical parameters of the industrial and model tank furnace were compared using the Wolf-Resnick complex flow model and MARMOD mathematical program.

Ein klassisches mathematisches Modell zur Bewertung der Fließdynamik von geschmolzenem Glas in Wannenöfen und gemäß einem komplexem Fließmodell wurde analysiert. Die Untersuchungen eines Wannenofens durch die Impuls-Antwortmethode wurden mit Hilfe des Radionuklids ^{140}La ausgeführt. Die damit erhaltenen technologischen und dynamischen Parameter eines industriellen und eines Modell-Wannenofens wurden unter Verwendung des komplexen Wolf-Resnick-Fließmodells und des MARMOD-Programms miteinander verglichen.

Keywords

dynamic function studies; flow models; furnaces; glass; lanthanum 140; mathematical models; radioisotopes; tracer techniques

1. Introduction

Tank furnace is a continuous operation open chemical reactor. Because of complex character of the molten glass flow within the tank, it proves itself rather difficult object for mathematical interpretation.

During the melting process of an oxide batch, the flow of material is affected by various interferences. As the result of the flow disturbances, the track and moment velocity of each particle is different. Thus the knowledge of the particle velocity distribution would be absolutely necessary for determining the real flow conditions in a unit. Experimental determination of the velocity distribution is impossible and therefore the flow pattern of a liquid stream enabling to perform engineering calculations should be found.

A very convenient characteristics constitutes the Residence Time Distribution determined by means of widely known impulse-response method. The principle of the method consists in introducing certain interference, e.g. radioactive tracer, and analysing the effects. As a result of such a procedure some parameters describing the flow pattern in an investigated object can be determined.

Necessary technological conditions of the glass melting process and physico-chemical changes occuring in the course of the process run evoke various flow disturbances. Also some unintended changes in the internal reactor geometry introduced by forming dead spaces, construction irregularities such as damages and constrictions of the furnace lining strongly affect the flow conditions.

An influence of axial and transversial convection currents and the currents stimulated by charging the row batch and draining the furnace by glass-drawing machines should also be taken into account.

[1]) Address: 05-400 Otwock-Świerk, Poland

Diffusion equalizing of concentrations plays also an important role. Surface currents cause by-passing of certain fraction of the total mass stream thus a part of material is not homogenized. Besides, the back-mixing currents and whirls caused by contraction of the effective volume of the reactor should be considered.

Fig. 1 shows hypothetic distribution of currents within the tank furnace. Possible ways of interpretation of the presented currents pattern have been discussed in a paper by *R. Malik* and *K. G. Kruszewski* [1].

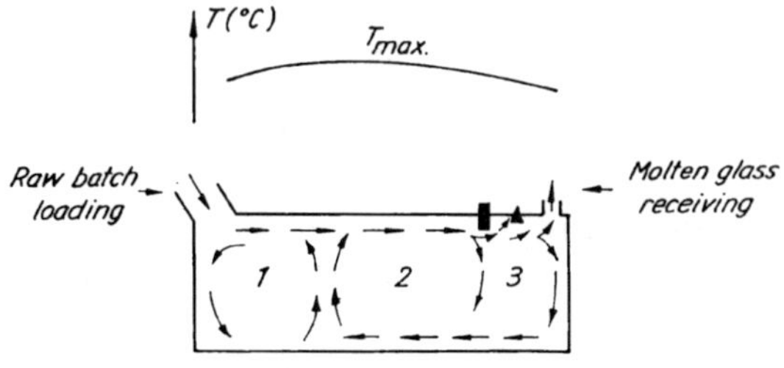

Fig. 1. Hypothetic distribution of currents within the tank furnace
1 – currents stimulated by loading the raw batch; 2 – downdraft currents (maximum temperature); 3 – currents stimulated by receiving the molten glass

Methodology of examining the tank furnaces using the impulse-response method was described by *A. R. Cooper* [2] and *L. Unger* [3] as far as back the 60 s. Results obtained by those authors bear merely descriptive and qualitative character and the experimental data do not reflect the dynamic flow pattern.

W. Tuszyński [4] has suggested a mathematical description of the dynamics of molten glass flow in a tank furnace. His result, however, arouse some doubts of technological nature.

It has been assumed that in the investigated reactor the molten glass flows according to the plug flow pattern while certain fraction of the total volume forms dead spaces in which the mass exchange processes cease. The author employs the equation describing the course of tracer concentration changes as a time function in order to determine the "β" parameter, which characterizes the process of tracer removal from the reactor:

$$N(t) = N_0(1 - e^{\alpha t_1}) \cdot e^{-\beta t_2} \cdot e^{-\lambda t_3} \tag{1}$$

Plotting the $N(t)$ response function in a logarythmic scale one can observe the region in which predominates the decrease of tracer concentration caused by uptaking the molten glass out of the furnace. This phemonenon is expressed by another parameter: a half-period of removal of the labelled mass, marked by T_w:

$$T_w = \frac{\ln 2}{\beta}$$

Unfortunately, using those parameters for the flow dynamics description frequently results in rather doubtful figures (e.g. contribution of dead spaces) and finally, the conclusion helpful for technological optimization cannot be drawn.

The above becomes quite obvious considering the fact that application of such a mathematical description requires adopting some physically impossible and mutually contradictory assumptions, such as instantaneous homogenization of the batch components, undisturbed plug-flow (no backmixing currents, absence of short-circuiting). These assumptions do not represent the real process conditions and never occur in practice as simply physically impossible.

Also a purely theoretical assumption concerning instantaneous perfect mixing remains in obvious contradiction to the plug-flow pattern adapted in the mentioned paper.

2. Method of identification

A theoretical description of the molten glass flow dynamics regarding all process and interferences actually occuring in the process of glass melting can be obtained on the grounds of the Residence Time Distribution functions of all particles of the flowing stream. For such a procedure adoption of certain mathematical model becomes necessary

Diffusion models cannot be employed in case of molten glass flow. Such models usually treat the reactor as a quasi-homogenic one, what means absolute absence of dead spaces or privileged streams [5, 6].

To some extent a simple chamber-model characterizes the molten glass flow dynamics as it bases on the assumption of similarity between mixing of a liquid in a single tube flow reactor and in a series of tanks being subsequent steps of perfect mixing [7, 8]. However the further assumption adopted in the chamber-model theory is less acceptable: there is no mixing and whatsoever changes of concentration between the elements of the model and the mass transfer from one step to another is instantaneous.

Real flow of molten glass in a tank furnace can be described only by means of more complex flow model.

The reactor must be considered as a set of regions of different flow character mutually interacting and providing complex multiparameter flow models [9].

Among the numerous published mathematical models of flow dynamics [8]–[11] the most adequate for glass technology purposes is the one proposed by *Wolf* and *Resnick* [12] Fig. 2 presents the block-diagram of this model:

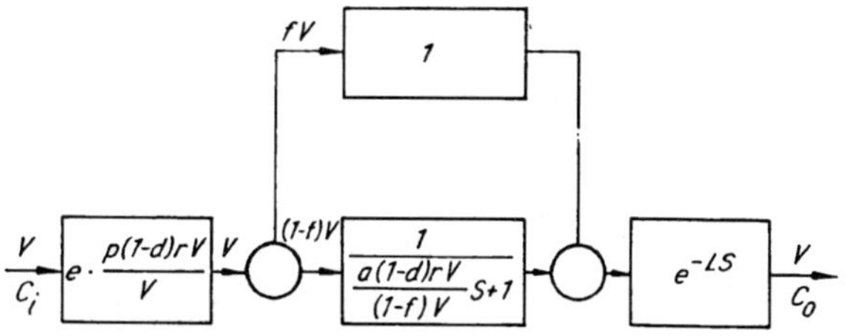

Fig. 2.

According to the model being discussed, the system can be described by the real mean residence time: rQ, where r is an error coefficient. The d fraction of the total system volume forms a dead space, thus $(1 - d)$ being an effective volume of the tank participating in direct flow of the molten glass.

The p coefficient expresses the part of the effective volume in which the plug-flow occurs; therefore if $p = 0$ the flow can be described by the perfect mixing pattern only; while for $p = 1$ the plug-flow pattern is applicable exclusively.

It has also been assumed that $1 - f$ expresses the part of the mass stream flowing in perfectly mixed forms and f is the fraction of the stream flowing at infinite velocity (by-pass). The stream described in such a way is also affected by the time-delay or lag, L. The lag can be of a distance-velocity type in a sampling system or any other delay influencing the response characteristics. Beside, the model takes into account the time delay between the impulse and response (expressed as m) as well as the part of the effective volume (expressed as a) in which perfect mixing occurs exclusively.

In the flow pattern developed in the above marmer, the function of tracer concentration changes at the reactor outlet is given in the following form:

$$F(t) = 1 - \exp\left\{-\frac{1-f}{a \cdot r(1-a)\,\theta} \times \left[t - L\frac{p(1-d)\,\theta}{1-f} + m \cdot a(1-d)\,\theta\right]\right\} \quad (2)$$

Another general approach proposed by the authors [12] gives the following form of the concentration changes function at the reactor outlet:

$$F(t) = 1 - \exp\left[-\eta\frac{t-\varepsilon}{\theta}\right]; \text{ for } t \geqq \varepsilon$$

while

$$F(t) = 0 \quad \text{for} \quad 0 < t < \varepsilon \quad (3)$$

The physical significance of the ε and η coefficients used in the equation (3) can be explained as follows:

η – indicates the extent of the real mixing of a liquid in the tank. For the perfect mixing pattern $\eta = 1$, while for the plug flow $\eta \to \infty$. Any deviation of η value from 1 implies a non-ideal mixing pattern;

ε – expressed in time units is a measure of the phase shift of the system response to tracer injection impulse. In case when perfect mixing occurs instantaneously $\varepsilon = 0$, while for the plug flow $\varepsilon = \theta$.
In most real cases of non-ideal mixing ε values can be higher or lower than θ.

Tab. 1. shows the η and ε/θ coefficients' values for different flow conditions.

Tab. 1. η and ε/θ values for different flow patterns

No	Flow pattern	η	ε/θ
1.	Perfect mixing	1	0
2.	Perfect mixing with plug-flow	>1	>0
3.	Perfect mixing with dead space	>1	0
4.	Perfect mixing with short-circuiting	<1	<0
5.	Perfect mixing with determined error of θ	$\gtrless 1$	0
6.	Perfect mixing with delayed reaction	1	>0

In the model proposed, all the dynamics phenomena taking place in the melting and clarifying parts of the tank furnace are considered.

In case of tank furnaces applied in the glass industry the adopted model can be usually restricted to merely two parameters, as the volume of molten glass phase and intensity of draining have constant values, what permits the exact theoretical mean residence time to be calculated. Basing upon the $N(t)$ function we can estimate the contribution of by-pass in the total mass-stream as well as the dead space function on the grounds of mean residence time.

Possible time delay arising at the stage of sampling and radiometric measurements has no significant importance regarding very low velocity of the molten glass stream (< 1 m/h).

Also the time in which the product passes through annealing furnace is always exactly known and necessary correction is introduced in the calculations.

3. Verification of the adopted model

A series of model investigations have been carried out to verify the rightfulness of the mathematical model employed for description of flow dynamics.

The model of the tank furnace is designed in 1:20 scale and built of plexiglass. Its construction is based on the largest in Poland type of furnace, provided with an automatic feeding system which loads accurately weighed raw oxide batch. This type of furnace is heated by natural gas burners. The final product is window glass, the sheets being made by six glass-drawing machines.

Simple representation of relatively high temperature of molten glass (1450 °C) in a model system is impossible and therefore the only parameters of the model liquid to be maintained are its viscosity and the flow intensity.

Viscosity of the molten glass at the temperature of 1450 °C amounts to approx. 100 P. Viscosity of the starch syrup used as a model liquid is similar at 20 °C and amounts to 80 P.

Character of flow of mass stream in a tank furnace is defined by the *Reynold*'s number, assuming that the flow within the furnace can be approximated by the flow through a tube. Then:

$$R = \frac{u \cdot D}{\gamma} \quad (3)$$

where: D – equivalent diameter
u – linear flow velocity
γ – kinematic viscosity of the molten glass

The originally adopted criterion of representativity was the constancy of the linear flow velocity, i.e. the following condition should be fulfilled: $U_{\text{furnace}} = U_{\text{model}} = \text{const.}$

Reynold's number and volumetric flow intensity were calculated from the experimental data. The verification criterion defined in such a way does not, however, prove the similarity of the flow character ($Re_f = Re_m$). As the similarity of the flow character is absolutely necessary for comparison between real model units and for estimation of the operation properties, another condition has been regarded as significant: $Re_f = Re_m$ thus:

$$\frac{U_f \cdot D_f \cdot \varrho_f}{\mu_f} = \frac{u_m \cdot D_m \cdot \varrho_m}{\mu_m} \quad (5)$$

where: ϱ_f, ϱ_m – specific gravity of molten glass and model liquid, respectively
μ_f, μ_m – dynamic viscosity of respective liquids

Values of the volumetric flow intensity obtained on the grounds of relationship (5) remain in good accordance with the velocity determined in the course of investigations carried out on the industrial furnace and allow to determine theoretical residence time of molten glass in the tank furnace.

The model investigations were planned with respect to the technological conditions the most frequently observed in industrial furnaces.

During the first series of experiments, uniform work of all the glass-drawing machines without any shutdown was assumed. A constant level of the model liquid in the tank was also maintained. The next variant represented the situation when one of the frontal machines was out of operation while the total uptake remained unchanged. The third cycle of experiments dealth with the situation when one of the side machines was stopped while the remaining technological parameters were maintained at a constant level.

^{24}Na radioisotope was used for the model investigations. The total activity used in each series was approx. 37 MBq. Sampling was carried on from the surface layer of liquid including melting part, skimming pockets, constrictions and drawing pot (Fig. 3).

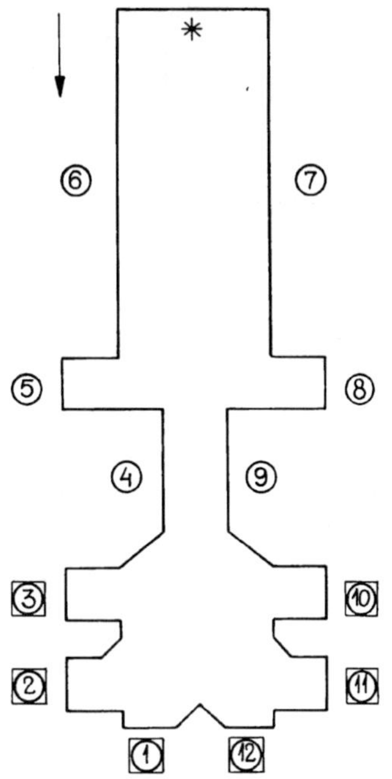

Fig. 3. Scheme of the model tank furnace and arrangement of measuring points for determination of tracer concentration changes as a time function

4 – 9 discrete measurement (sampling); 1 – 3 and 10 – 12 – continuous and discrete measurement; ✳ – tracer injection point

Gravitional feeding was applied. Sampling was repeatedly performed in all chosen points at one-hour intervals.

Burners in the model tank were replaced by IR radiators. Number of radiators was exactly the same as the number of burners in actual furnace: 7 along each side-wall of the tank. They were alternatively switched on each side every 15 minutes to stimulate the flame reversion.

The tracer activity measured was referred to 1 g of model liquid. Typical electronic system: HT supplier-pulse counter ("Standard"-type) was used for radiometric testing of the samples. Detection of radiation was performed by means of scintillation probe provided with NaI(Tl) crystal.

All calculations of the model parameters were carried out using a digital computer operating according to modified MARMOD program. The program is based on the least-square method; in the course of data processing the program minimizes the sum of deviations of experimental data from the parameters' values of selected mathematical model.

Technological parameters of operation for both: industrial and model furnaces are presented in the Tab. 2. Tab. 3 gives a comparison of dynamic parameters of both units, calculated using the mentioned procedure.

Previously adopted model, assuming ideal flow pattern for the process identyfication [2–4] does not provide full representation of actual phenomena taking place in real tank furnace during the melting process and molten glass transport.

Corrected method of mathematical interpretation of the experimental data based on the weighted moments method

Tab. 2. Comparison technological parameters of industrial and model bath

No	Name of parameters	Dimension (measurement)	Bath industrial	Bath model
1.	Total liquid measure	dm^3	$544 \cdot 10^3$	68
2.	Density	kg/dm^3	2·5	1·33
3.	Work linear length of aggregate	cm	4900	245
4.	Time of buffer	min	480	46
5.	Volume of buffer	dm^3	$27{\cdot}76 \cdot 10^3$	23·55
6.	Speed of extraction	dm^3/min	55	0·51
7.	Translocation max. speed	dm/min	1·02	0·53
8.	Translocation mean speed of liquid	dm/min	0·02	0·18
9.	Theoretical time of extraction the whole of liquid	min	9420	133
10.	Mean time of distribution	min	240	108
11.	Active volume	dm^3	$416 \cdot 10^3$	55·08
12.	Volume area of stagnancy	dm^3	$128{\cdot}3 \cdot 10^3$	12·92
13.	Aggregate advantage factor	%	76·54	81
14.	Mean time variant of distribution	min^2	$3139 \cdot 10^2$	3534
15.	Undimensional standard deviation			0,303
16.	Alternative labelling elimination	min	6030	131
17.	Pecklet's number		5·5	5·37
18.	Number of stage ideal mixing		2·75	3·30
19.	Real mixing stage	n –	2·23	3·26
20.	Phase shift factor	E –	0·48	0·65
21.	Reaction decelerate for disturbance system	min	3456	71

The method of parameters calculation are given on INR internal description [13].

enables to determine accurately the mean residence time and some other parameters of tank furnaces commonly used in glass industry.

Developing the mathematical model for the molten glass flow pattern gives possibility for estimation of the dynamic phenomena occuring in the melting process and mass transport within the furnace.

The model proposed remains in good accordance with the experimental response function.

Certain discrepancy between the dynamic function curves (Fig. 4), for the time variable close to $2t$ can be explained on the grounds of insufficient accuracy of the discrete mathematical interpolation of the curve.

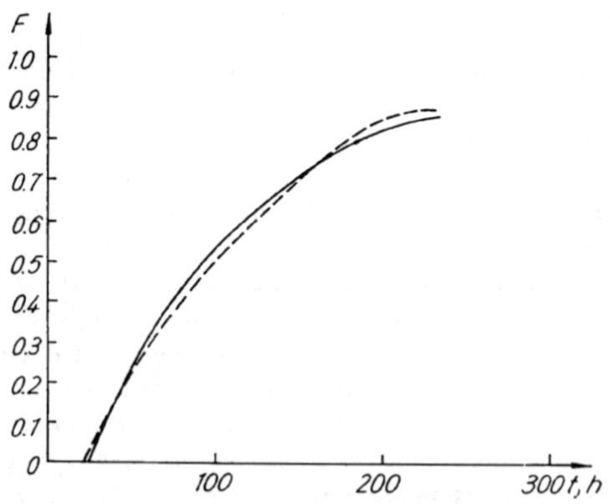

Fig. 4. Dynamic characteristics of the molten glass flow

Comparison of dynamic characteristics of the molten glass flow in a commercial tank-furnace for window glass production (6 glass-drawing machines)

- - - - – Experimental curve; ——— – Theoretical curve calculated on the grounds of *Wolf-Resnick* model

Tab. 3. Comparison dynamic parameters calculation of industrial and model bath work in use with complicated mathematical model (progr. MARMOD)

Aggregate characteristic	t_L	$\bar{t}$	d	f	m	r	n	E	a	p	$a+p$
Six-machines bath with constriction for window glass production with capacity $544 \cdot 10^3$ dm³	157	120	0·24	0·02	0·02	0·90	2·23	71	0·836	0·164	1
Model bath with capacity 68 dm³	2·22	1·80	0·130	0	0	0·96	3·26	3453	0·370	0·630	1

Relatively low contribution of perfect mixing ($a = 37\%$, Tab. 3) in the model tank is obviously caused by absence of thermal currents, while in actual melting process they play a significant role in mixing and clarifying of the molten glass (Fig. 1).

4. Concluding remarks

1) The *Wolf-Resnick* model has been proposed for identification of the molten glass flow process in industrial tank furnaces.
2) Phenomenological approach to glass melting process, based upon the *Wolf-Resnick*'s function, gives practical solution fully applicable for purposes of industrial process optimization.
3) The above statements have been fully proved by the results of experiments carried out on the model tank-furnace.
4) The proposed method has been applied for determination of the melting process parameters in the industrial tank furnace. The results obtained constitute strong affirmation for the entire concept and are a subject of separate paper.

Received April 3, 1985

References

[1] *R. Malik, K. G. Kruszewski*, Postępy Techniki Jądrowei **14** (1970) 821
[2] *A. R. Cooper*, Jr., Journ. of the Amer. Ceram. Soc. **12** (1960) 93
[3] *L. Unger*, Glastechn. Berichte **33** (1960) 416
[4] *W. Tuszynski*, Szkło i Ceramika **10** (1963) 261
[5] *O. Levenspiel, W. K. Smith*, Chem. Eng. Sci. **6** (1957) 227
[6] *V. Ananthakrishnan*, A.I.Ch.E. Journal **11** (1965) 1063
[7] *E. Kirschbaum*, Forsch. Geb. Ing. **135** (1935) 245
[8] *W. W. Dilman*, Chim. Promyszl. **40** (1964) 511
[9] *O. Levenspiel*, Can. J. Chem. Eng. **40** (1962) 113
[10] *I. A. Aleksandrow, W. G. Wybornow*, Teor. Osnow. Chim. Techn. **5** (1971) 339
[11] *R. L. Solomon, J. L. Hudson*, A.I.Ch.E. Journal **17** (1971) 371, 379
[12] *D. Wolf, W. Resnick*, Ind. Eng. Chem. Fundamentals. (1963) 287
[13] *K. G. Kruszewski* et al., INR Internal Report., 21, XVI (1978)

Isotopenpraxis 22 (1986) 2, S. 53 bis 57

Solvent Extraction Technology of $^{99}Mo-^{99m}Tc$ Generator System 1. An Indian Experience: Process Design Considerations

O. P. D. Noronha

(Radiation Medicine Centre, Bhabha Atomic Research Centre, c/o Tata Memorial Hospital Annexe, Parel, Bombay)[1])

The solvent extraction method of separating and purifying $Na^{99m}TcO_4$ from low specific activity ^{99}Mo is rarely employed in developed countries, whereas in India $Na^{99m}TcO_4$ has been separated, purified and used in an injectable form for the past decade using low specific activity ^{99}Mo and an indigenous separator system. The entire operation was previously carried out by a manual method. In order to decrease the radiation burden to the personnel performing these operations the process assembly was redesigned to incorporate additional safeguards against radiation. The paper decribes the modifications introduced and our experiences with this 'semi-automated' system.

In entwickelten Ländern wird das Verfahren der Lösungsmittelextraktion zur Trennung und Reinigung von $Na^{99m}TcO_4$ aus ^{99}Mo mit geringer spezifischer Aktivität kaum angewandt. In Indien z.B. wurde in den letzten 10 Jahren das $Na^{99m}TcO_4$ in einer injizierbaren Form unter Verwendung von ^{99}Mo mit geringer spezifischer Aktivität und eines einheimischen Trennsystems getrennt, gereinigt und eingesetzt. Die gesamte Operation wurde früher von Hand durchgeführt. Um die Strahlungsbelastung für das Personal, das diese Operationen ausführte, zu verringern, wurde die Produktionsvorrichtung umgebaut, indem zusätzliche Sicherheitsvorkehrungen gegen Strahlung vorgesehen wurden. Die Arbeit beschreibt die durchgeführten Veränderungen und unsere Erfahrungen mit diesem ,,halbautomatisierten'' System.

Keywords

molybdenum 99; nuclear medicine; radioisotope generators; solvent extraction; technetium 99m; pertechnetates

Introduction

The radionuclide ^{99m}Tc is the work-horse of present day nuclear medicine. More than 70% or nuclear medicine investigations are being performed with ^{99m}Tc-radiopharmaceuticals [1, 2]. Sodium pertechnetate, ^{99m}Tc, in addition to being a radiopharmaceutical per se, is also the basic raw material for the synthesis of a variety of other ^{99m}Tc-derivative agents. This material is generally obtained in an injectable form by eluting sterile column (alumina) chromatographic type of devices containing $^{99}Mo-^{99m}Tc$ ge-

[1]) Address: Bombay 400012, India

nerator systems. The high specific activity ^{99}Mo that is loaded onto such columns is either fissionproduced or obtained by neutron-irradiation of enriched ^{98}Mo in high flux reactors [3, 4]. Recently, attempts have been made to fabricate zirconium molybdate (^{99}Mo) gel-type column chromatographic systems employing low specific activity ^{99}Mo [5]. In addition to these methods $Na^{99m}TcO_4$ has also been prepared by sublimation and solvent extraction (S.E.) techniques [3, 4]. However these two latter methods have not found wide acceptance with individual user (hospital radiopharmacy) laboratories. Their use has been restricted to a few selected central/regional radiopharmacy dispensing laboratories or mostly relegated to pilot studies. We, in India, have relied upon low specific activity ^{99}Mo obtained by irradiating natural (not enriched) molybdenum in low-power reactors as our source for ^{99m}Tc for more than a decade. The $Na^{99m}TcO_4$ injectable was prepared over this period by separating $^{99m}TcO_4^-$ from $^{99}MoO_4^{2-}$ by the solvent (methyl ethyl ketone) extraction method in an indigenously designed extractor system [6]. The various processes-separation, purification and further processing to convert $Na^{99m}TcO_4$ into a parenteral form – were carried out by a manual procedure. This report summarizes some of the improvements that have been introduced in the process design (by way of arrangement of assemblies, use of remote-handling and shielding contraptions) in order to reduce the radiation exposures to the personnel; and our experiences with this system.

Materials and methods

Sodium molybdate (^{99}Mo) was received from the manufacturer (Isotope Group, B.A.R.C., Bombay 400085, India) as an aqueous solution in 5 N NaOH. It was consigned in four (lead shielded) 10 or 15 ml capacity glass vials. Each vial contained $\sim$ 200 mCi and had a radioactive concentration of 2·5 – 39·0 mCi/ml. The specific activity of ^{99}Mo varied from 9·8 – 300·0 mCi/g of Mo on the first day of use. Generally the contents of two vials ($\sim$ 400 mCi) were used (solvent extracted for ^{99m}Tc) on the first day. The contents of the other two vials were utilised (i.e. added on to the previous day's pooled ^{99}Mo source transferred from the ^{99}Mo storage receptable vial into the extractor) on the second and third day respectively. The radiochemical separation was affected using an indigeneously designed lead-shielded all-glass extractor of *Mani* and *Narasimhan* [6], and following the stipulated instructions. The various radiochemical operations of S.E. separation, purification and repurification, were carried out in an efficiently exhausted fume cupboard. The transfer of ^{99}Mo solution from the consigned vial containing ^{99}Mo or the ^{99}Mo source receptacle vial (vide infra) and mixing of the organic (methyl ethyl ketone (MEK)) and aqueous-alkaline phases were effected by using an electrically powered miniature vacuum/pressure pump (Millipore Corpn., Bedford, Mass., U.S.A.). The pump was connected to the extractor through a threeway stopcock and a trap. A shielded 100 ml. capacity vial positioned behind the shielded extractor served as the receptacle vial for storing the pooled ^{99}Mo solutions after the $^{99m}TcO_4^-$ extraction. The stored solution was used on the next day as the $^{99}Mo-^{99m}Tc$ source. The extraction was generally carried out once a day, and the ^{99}Mo source used for $\sim$ 6 days (a week). The different remote-controlled or lead-shielded assemblies or devices (eg. lead shielded beakers, a short glass column containing basic alumina – 1·5 cm × (1·5 – 2·0) cm dia. – on a medium porosity frit disc supported on an *Erlenmeyer* flask for purification of $^{99m}TcO_4^-$ in MEK, or a small thermostatically controlled hot-plate or water bath, etc.), as well as glass mirrors (placed at an incline at the rear) were positioned in the fume cupboard before the start of the operations. The radiopharmacy operations were carried out in a separate but adjacent lead-shielded enclosure alongwith a table-top shield of lead and lead-glass (Atomic Products Corpn., Center Moriches, N.Y., U.S.A.).

Results and discussion

The preparation of $Na^{99m}TcO_4$ injectable by the manual S.E. method involves a number of steps. Also it involves the handling of large amounts of both the nuclides ($^{99}Mo + ^{99m}Tc$) and ^{99m}Tc, as open and wet sources within the precincts of a (sometimes less well-equipped) user hospital radiopharmacy laboratory. Some of the procedures involved in a manually-operated S.E. process are schematically illustrated in Fig. 1. These include both the radiochemical and radiopharmacy operations. The former include (I) the radioactive assay of the consigned vial for total activity, desealing of the vial and the transfer of ($^{99}Mo + ^{99m}Tc$) contents as well as the alkaline washings into the extractor, (II) the mixing of the two phases – aqueous alkaline (5 N NaOH) containing $^{99}Mo + ^{99m}Tc$ with the organic (MEK) phase, (III) the separation of the two phases, and removal of the ^{99}Mo depleted of Tc into the storage receptacle vial, (IV) further scrubbing the $^{99m}TcO_4^-$ in MEK with 5 N NaOH and removal of waste aqueous-alkaline phase, (V) passing the $^{99m}TcO_4^-$ in MEK over a short length of alumina column, and washing the column again with MEK to remove any adsorbed $^{99m}TcO_4^-$, (VI) evaporation of the solvent MEK, and (VII) leaching or dissolution of $^{99m}TcO_4^-$ with physiological saline. The radiopharmacy processes involve (VIII) membrane (0·22 μm) filtration of aqueous solution of $^{99m}TcO_4^-$ in saline, (IX) radioactive assay for ^{99m}Tc, and dispensing, if required, and (X) formulation and radioactive assay (for ^{99m}Tc) of the different ^{99m}Tc-radiopharmaceuticals. It is evident that performing these multiple 'hot' operations without adequate safeguards could result in personal exposures because of the radiation field.

The lead-shielded (2·5 cm thick) all-glass solvent extractor assembly designed by *Mani* and *Narasimhan* [6] for the separation and the first stage purification of $^{99m}TcO_4^-$ is an efficient system, and has been routinely used by us for the past decade. The process has also resulted in good yields and product purity. However, the shielded extractor provides a modicum of radiation protection only during the separation stage of the process. In fact the various manually operated processes can only be safely performed by skilled operators while handling low activity levels of $^{99}Mo-^{99m}Tc$. Even this calls for a fair degree of expertise in manipulating the various procedures if the radiation exposures are to be kept under check or as low as reasonably achievable (ALARA). In order to operate the process to suit our requirements (handling upto 600 mCi amounts of $^{99}Mo + ^{99m}Tc$ per day) we redesigned/realigned the various assemblies to incorporate a greater degree of radiation protection. The Fig. 2 illustrates the general lay out of the various assemblies, improvised contraptions, semi-automated devices and other modifications that were used/introduced for carrying out the various radiochemical and radiopharmacy operations. These two operations were

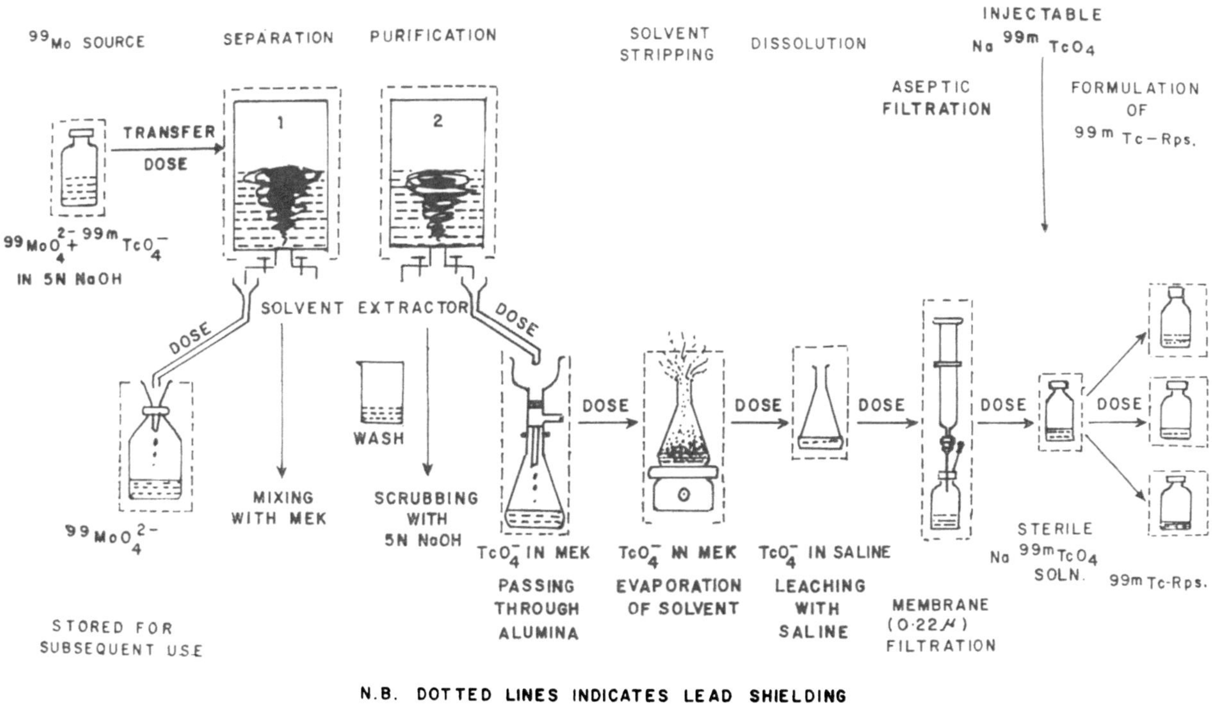

Fig. 1. The important processes involved in the preparation of $Na^{99m}TcO_4$ injectable and formulation of other ^{99m}Tc-radiopharmaceuticals (Rps.)

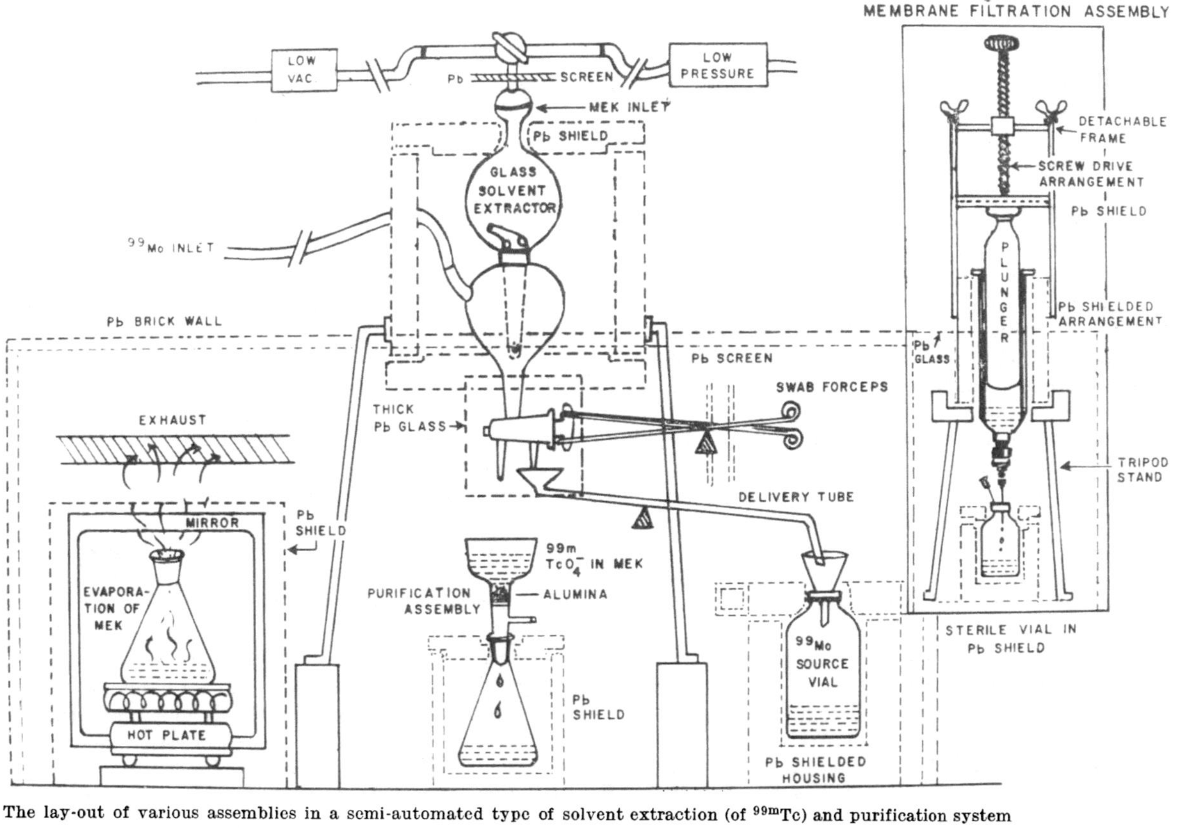

Fig. 2. The lay-out of various assemblies in a semi-automated type of solvent extraction (of ^{99m}Tc) and purification system
Inset shows the lead-shielded membrane filtration assembly

performed in separate enclosures to avoid cross-contamination and in keeping with Good Manufacturing Practices (GMP).

To obviate or considerably reduce the radiation dose that could be expected during the preparation of $Na^{99m}TcO_4$ by the manual mode the process assembly was redesigned to incorporate additional elements of radiation protection; viz. lead shielding, increasing distance and reducing the time factor of the operations. The same was also achieved by (semi) automating some of the 'hot' processes. The operations of (relatively) greater radiation dose and therefore risk were those involving steps (I)−(III); i.e. handling of ($^{99}Mo + ^{99m}Tc$) rather than ^{99m}Tc solutions. And therefore, these processes were semi-automated to safeguard the operators. The transfer of ($^{99}Mo + ^{99m}Tc$) solutions either from the consigned vial or storage receptacle vial into the shielded extractor was effected by employing low vacuum. The mixing of the two phases-aqueous alkaline and organic solvent MEK-was achieved by passing a stream of air from a low pressure pump. Both these operations could be sequentially performed from a distance by actuating an electrically operated low pressure cum vacuum pump (Millipore Corpn.). The ^{99}Mo solution (depleted of ^{99m}Tc) was removed by operating the teflon stopcock. This was done by using a pair of long-handled swab forceps, fixed to the stopcock (before the start of the operations) from behind a lead screen. The ^{99}Mo solution was drained off into its container (^{99}Mo source receptacle vial) by means of a bent delivery tube held in place by means of a clip attached to a short support stand. The vial for or containing ^{99}Mo was not moved around, but kept in a convenient fixed position (behind the extractor) inside a specially fabricated extra-thick lead-shield housing. It was also covered on the top with a snugly fitting lead cover provided with a central opening (for entry for the ^{99}Mo inlet tube or the end of the delivery tube). Similar operations were also performed for step (IV) using 5 N NaOH as the scrubbing agent. The waste was collected in a beaker fully covered with a tight-fitting lead shielded or else emptied into a shielded carboy (for waste) by means of a delivery tube. All other operations were performed using lead shielded assemblies and an assorted variety of extralong handled (remote-controlled) devices − pair of (specially fabricated) tongs/forceps, radiarm, grip tongs (Atomic Products Corpn.), etc. The fume-cupboard was shielded on all sides to a height of 90 cm with 0·6 cm thick lead sheeting. In addition it was provided with a 30·0 cm tall lead-brick wall (5·0 cm deep) on the front side. A 10·0 cm thick lead-glass was placed in front of the constricted portion of the extractor near the stopcock end in order to view the two liquid phases during the phase separation step. Most of the assemblies were positioned in the fume-cupboard before the start of the operations. This helped to avoid unnecessary exposures during the 'hot' operations. Also sufficient perambulatory space (> 2·0 m) was provided in front of the fume cupboard in order to maintain safe distance from the vicinity of the 'hot' operational area.

Protection against radiation while handling ^{99m}Tc nuclide (steps V−X) was also achieved by following the abovementioned radiation attentuation procedures. (The 'hot' sources were never touched with hands.) The repurification of $^{99m}TcO_4^-$ in MEK (step V) was carried out by positioning the purification assembly (made up of a short glass column containing alumina supported on a frit disc and kept in a lead-shielded *Erlenmeyer* flask) just below

Fig. 3. A general view of the arrangement of assemblies in the fume cupboard (mainly for radiochemical operations) and in the adjacent enclosure (for radiopharmacy operations)

the ^{99m}Tc outlet of the stopcock, or using a bent delivery tube originating from the stopcock into the system. The other operations (VI−VII)) were carried out by using extra long-handled tongs or by performing the operations behind shielded assemblies or portable lead screens. (For example, the steps VI and VII were carried out on a hot plate fully lead shielded in front − except for a narrow slit − and partially open at the rear end for viewing the operations in the mirror placed at the rear end of the fume cupboard). Since the manual handling of the membrane (0·22 μm) filtration of $Na^{99m}TcO_4$ (i.e. converting the radiochemical into a sterile radiopharmaceutical form) could result in significant radiation exposures to fingers and hands, this operation was performed using a remote-handled shielded contraption. The shielded membrane filtration assembly is shown in the Fig. 2 (inset). The Fig. 3 gives a general view of the arrangement of assemblies in the fume cupboard and in the juxtaposed enclosure for carrying out the membrane filtration operations in our laboratory. The radiation exposures during dispensing and formulation procedures (IX−X) were drastically reduced by employing conventional shielding techniques, e.g. enclosing the vial containing $Na^{99m}TcO_4$ in a lead pot (collection-elution vial shield Amersham International plc, U.K.), or a locally fabricated lead pot with a slip-on lid provided with a small opening at the centre for inserting the needle of the syringe, and fastened to the body of the pot by clamps and screws. The other radiation protection devices that were used in the operations were lead-glass, P.V.C./neoprene gloves loaded with lead (Elastotechnic, Paris, France; Atomic Products Corpn.), Radiglasses for eyes, lead-lines aprons, lead shielded syringes (with lead-glass windows), etc.

The prime consideration for choosing a particular technology (for the routine separation and preparation of $Na^{99m}TcO_4$) in a developing country is the cost as well as other socio-economic factors such as the ready availability of an indigeneously developed technology. The other major considerations are the yield of $Na^{99m}TcO_4$, its purity, the radiation dose encountered in operating the various processes with its accompanying radiation exposures to operating personnel. Equally important factors are the level of expertise required in carrying out the operations, the relative ease or complexity of these operations for routine use, the time required for completing the operations, and the consistency and reliability of the process and the end product. We have adopted the S.E. method

Mitteilungen / Information

IAEA: DIE GENERALKONFERENZ DER IAEA SEPT.1985
Die 29. ordentliche Tagung der Generalkonferenz der IAEA wurde am 23.9.85 in der Hofburg eroeffnet. Es nahmen mehr als 90 von 112 Mitgliedstaaten der Organisation an der Konferenz teil. Zum Praesidenten der Konferenz wurde Herr Adonit MANOUAN, Botschafter der Elfenbeinkueste, gewaehlt; Vorsitzender des Plenarausschusses war Herr Hans SCHEEL, DDR. Auf der Konferenz wurde das Budget der IAEA fuer 1986 in Hoehe von fast US$ 100 Millionen bewilligt und ueber die freiwilligen Beitraege zum Fonds fuer Technische Hilfe und Zusammenarbeit beraten.
(IAEA-Pressekommunique XXIX/1)

Kurznachrichten / Newlets

KANADA 1984 GROESSTER U-PRODUZENT DER WELT
Laut UIC-Newsletter der australischen Uranium Information Centre Limited (UIC) produzierte die U-Minen der kanadischen Provinzen Saskatchewan und Ontario 1984 rund 11.170 t U-Konzentrat. Kanada war damit im vergangenen Jahr der Welt groesster U-Produzent. Sein Anteil an der U-Produktion der westlichen Welt betrug 1984 laut derselben Quelle 29% gegenueber 19% im vorangegangenen Jahr. 15% der kanadischen U-Foerderung wurde zur Deckung des Inlandbedarfs verwendet. (SVA-Bulletin (1985)13)

Veröffentlichungen / Publications

STAATLICHES AMT FUER ATOMSICHERHEIT UND STRAHLENSCHUTZ

SAAS-325
Die aeussere berufliche Strahlenbelastung 1982 in der DDR - 17 S., 1985, W. Rothe

SAAS-326
Die aussere berufliche Strahlenbelastung 1983 in der DDR - 17 S., 1985, W. Rothe

SAAS-328
Empfehlungen zur hydrogeologischen Begutachtung der Standorte von Kernanlagen im Rahmen des Strahlenschutz-Genehmigungsverfahren - 56 S., 1985, Ch. Adam u.a.

SAAS-329
Besonderheiten des Brandschutzes in Kernkraftwerken - 12 S., 1985, W. Krueger

ATOMIC ENERGY OF CANADA

AECL-8647
Atomic Energy of Canada ltd. - Progress Report Health Sciences Division 1984 July 1-Dec.31/ 116 pp, 1985

AECL-8648
Progress Report Physics Division 1984 Juli 1- Dec. 31, 1985

AECL-8704
Progress Report Chemistry and Materials Division 1984 July 1-Dec.31, 116 pp, 1985

AECL-8379
Modification of radiation response in mammalian cells : the repair and fixation of X-ray-induced potentially lethal damage, 49 pp, 1985, E.I. Azzam, G.P. Raaphorst

AECL- 8380
Use of the food-chain model food III and the soil model SCEMR to assess irrigation as a biosphere pathway, 29 pp, 1985, S.C. Sheppard

KERNFORSCHUNGSANLAGE, KARLSRUHE

KFK-3883
Jahresbericht 1984 der Hauptabteilung Sicherheit, 257 S., 1985

KFK-3920
Experiments for training in nuclear and radiochemistry, 216 pp., 1985, S. Mobius

KFK-3909
Matrix correction for PIXE in biomedical samples, 23 pp., 1985, D. Heck, E. Rokita

GESELLSCHAFT FUER REAKTORSICHERHEIT MBH

GRS-84
Jahresbericht 1984, 85 S., 1985

RISØ NATIONAL LABORATORY-ROSKILDE

RISO-R-515
Dry Deposition of Urban Surfaces, 23 pp., J.Roed

RISÖ-R-520
Department of Energy Technology Annual Progress Report 1 Januar - 31 December 1984, 53 pp., B. Micheelsen, F. List

RISO-M-2410
Isotoplaboratoriet. 1959 - 1984, 42 pp., 1984

RISØ-M-2438
Dry deposition and resuspension of particulate matter in city environments, 36 pp., 1984, N.O. Jensen

RISØ-M-2454
Neutron Radiographic Techniques, Facilities and Applications, 26 pp., 1984, J.C. Domanus

RISO-M-2493
Atmospheric boundary modeling, 39pp., 1985, N.O. Jensen

RISØ-M-2474
Calculated shielding factors for selected European houses, 33 pp., 1984, P.J. Hedemann

TECHNICAL RESEARCH CENTRE OF FINLAND (Espoo)

VTT-357
Simulation of flame acceleration in unconfined vapour cloud explosions, 53 pp., L. Kjaldman, R. Huhtamen

Ausgewählte Beiträge / Recently published papers

ATOMNAJA ENERGIJA 59(1985)6 (all in Russian)

Problems of Chemical Analytical Monitoring in Atomic Power Facilities - L.N. Möskvin

Investigation ans Selection of New Extractors for the Removal of Actinide Elements - A.M.Rozen

A Mathematical Model of the Temperature Field Around Wells with Radioactive Wastes and Its Verification Under Field Conditions - E.G. Drozhko et al.

Passage of Primary Protons Through Shielding Characterizing Accidental Distribution of Matter - V.G. Mitrikas et al.

Application the Pade Approximation to Calculate Sub-Group Constants and Accounting for Doppler Effect in Analyzing Resonance Neutron Cross-Sections - S.A. Badikov et al.

Measurements of U-236 to U-235 Fission Cross-Section Ratios for 4 to 11 MeV Neutrons - A.A. Goverdovsky et al.

A Study of Lead Distribution in Rocks by the Radiography of Fission Fragment Tracks - V.P. Perelygin

Om Three-Dimensional Calculations of a Subcritical Heterogeneous Reactor with Neutron Source - V.M. Malofeyev

Effect of Finite Dimensions of a Moderator on the Characteristics of a Slow-Neutron Pulse-Source - N.I. Alekseyev et al.

Liquid Standard Gamma-Ray Sources - B. Ya. Shcherbakov

A Specialized Mass-Spectrometric Facility for Analyzing Aggressive Gas Mixtures - N.N. Bobrov-Egorov et al.

Equivalent Doses of Roentgen Radiation in Heterogeneous Human Phantom - V.I. Ivanov

J. of RADIOANAL. NUCL. CHEM. ART. 90(1985)2
Praeparation of 113 m In radiocolloid for liver scanning - M.A.A.Al-Janabi et al.

A radioreagent method for determination of traces of phosphorus in high-purity silicon - P.Y. Chen et al.

Statistical analysis of biodistriburion of 99mTc- colloid preparations - V. Jovanovic et al.

Simultaneous determination of rare earth elements by using substoichiometric separation of lanthanum - M. Katoh et al.

Extraction and neutron activation analysis of humic substances - K. Randle

Thorium ore analysis by X-rays in neutron activation analysis - P. Arikan et al.

Analytical expressions for the intensity of characteristic X-radiation and their relation to the measuring geometry of X-ray fluorescence analysis

Studies in heavy ion activation analysis, VII - C. Friedli et al.

Studies in heavy ion activation analysis, VIII - C. Friedli et al.

Determination of fluorine in geological samples using accelerator derived neutrons - K. Randle

Determination of trace and minor elements in coal: A comparison between instrumental photon and thermal neutron activation analysis - T. G. Pringle et al.

Provenance studies of the white marble of the cathedral of Como by neutron activation analysis and data reduction - M. Oddone et al.

Rapid dissolution of plutonium in soil by fusion with ammonium hydrogen sulfate followed by plutonium determination by ion exchange and alpha spectrometry - M. Yamamoto

Range calculation of alpha-recoil atoms in some minerals using LSS-theory - T. Hashimoto et al.

Radiochemistry and radiochemical separations - E. Bujdoso

J. of RADIOANAL. NUCL. CHEM. LET. 95(1985)1
The modified cryogenic adsorption method for the laboratory separation of Kr-85 from the atmosphere - L. Wilhelmova et al.

Study on the comparator method using substoichiometry I. Principle - M. Katoh et al.

J. of RADIOANAL. NUCL. CHEM. LET. 95(1985)2
Analysis of ancient iron objects with 14 MeV neutron activation analysis - V.N. Bhoraskar et al.

A rapid method for the determination of radionuclides in reactor coolant by ion exchange membranes - C.-C. Chan et al.

Simultaneous determination of U and Th by delayed fission neutron technique based on neutron generator - G.I. Khalil et al.

Determination of mercury by radiochemical displacement - B.V. Acharya et al.

Extractive radiometric procedure for the estimation of fluoride - A.D. Naidu et al.

J. of RADIOANAL. NUCL. CHEM. LET. 95(1985)3
Determination of trace elements in fish by instrumental neutron activation analysis - R.M. Awadallah et al.

Determination of carbon in electroplated gold layers by deuteron activation analysis - C. Vandecasteele et al.

Determination of arsenic, scandium, chromium, cobalt and nickel in asbestos by neutron activation analysis - D.K. Teherani

Radioluminescence quenching of toluene scintillators by methyl metacrylate - Z. Polacki

Production of electron-hole pairs in SiO_2 films - I. Petr

J. of RADIOANAL. NUCL. CHEM. LET. 95(1985)4
Decontamination of some radioactive effluents by means of ceramic filters - Al. Cecal

An improved titrimetric method for the determination of oxygen to metal ratio in uranium oxide - F. Khatoon et al.

The effect of hydrogen bromide on the methane production in the gamma radiolysis of the H_2S-CH_3Br system - I. Chrzascik et al.

Radiation induced defects on Si-SiO_2 interface - I. Petr

The multiphase kinetics of phosphate uptake by plant roots - I. Michalik

J. of RADIOANAL. NUCL. CHEM. LET. 95(1985)5
Determination of micro amounts of copper by radiochemical displacement - K. Pareedulla et al.

Some effects of temperature on fluorescence for pure toluene and liquid scintillator - Y. Homma et al.

Uranium and thorium analyses by delayed fission neutron counting technique using a small neutron generator - N.K. Mumba

J. of RADIOANAL. NUCL. CHEM. LET. 95(1985)6
The separation of Am-241 from bones by coprecipitation with bismuth phosphate - Z. Hoelgye

Influence of root structure on uptake and transport of radioactive phosphorus - I. Michalik

ATOMKERNENERGIE/KERNTECHNIK 47(1985)2
Die Masse-Energie-Beziehung und ihre Folgen. Zu einem zweifachen Jubilaeum - W. Hanle et al.

Heutiger Stand der Brennelementherstellung mit plutoniumhaltigem Brennstoff - K. Ennerst et al.

Betriebserfahrungen mit der Wiederaufarbeitungsanlage Karlsruhe - W. Schueller

Uranextraktionszyklus in der Anlage TEKO - K. Eiben et al.

Analyse stickstoffhaltiger Verbindungen in Purex-Prozessloesungen - D. Ertel

Ueber die Notwendigkeit der chemischen Mikrobereichsanalyse bei radioaktiven Nebenprodukten des Brennstoffkreislaufs - F. Baumgaertner

Auswahlkriterien fuer Denitrierungsverfahren in Brennelement-Wiederaufarbeitungsanlagen - S. Halaszovich et al.

Bildung und Verhalten brennbarer Gase bei

Stoerfaellen und schweren Unfaellen in Druckwasserreaktoren - K. Hassmann, H. Alsmeyer

Beherrschung von Natriumbraenden und Natrium-Wasser-Reaktionen in Brutreaktoren - K. Foerster et al.

Differentielle Gammastrahlenfluesse nach Durchgang von 6,13-MeV-Strahlung durch Schichten mit kleiner Ordnungszahl - J. Banai, G.B. Bishop

Radiolyse von Freon F 12 und seinen Gemischen mit Methan und Sauerstoff - W. Fuerst, H. Heusinger

Auswaschung von Schadstoffen in der Atmosphaere durch Schnee - W. Kuehn et al.

ATOM (1985)347

Increasing oil production from the North Sea - St. Hubbard

The DeBeNe programme - a history of collaboration - D. Grist

Research and Development for the UK PWR - D. Hicks

The future of the ESI in the UK - P.M.S. Jones

IAEA-Bulletin 27(1985)3

Nuclear power and the electronics revolution - L. Konstantinov et al.

Control and safety computers in CANDU power stations - R.S. Gilbert

Advanced computerized operator support systems in the FRG - W.E. Buettner

Simulation technology in operator training - T. Perkins

New concepts in control-room design - J. Furet

Robots for nuclear power plants - T. Moore

Remote and automation technologies: Review of Japan's experience - K. Uematsu

Reassessing radiation releases: A closer look at source term - M. Rosen, M. Jankowski

The United Nations at 40

Probabilistic safety assessment: Growing interest - L. Lederman

More evidence on PSA - M. Cullingford

Nuclear plant availability: Real achievements - J. Daglish

ATOMWIRTSCHAFT/ATOMTECHNIK 30(1985)10

Strahlen- und Umweltforschung in Neuherberg - Konzeption - Aufgaben - Perspektiven - H.W. Levi

Strahlenschutzorientierte Forschung in der GSF - U.H. Ehling u.a.

Forschung fuer die Endlagerung radioaktiver Abfaelle - K. Kuehn u.a.

Das Potential der deutschen Zulieferindustrie - H. Bonnenberg

Exportmaerkte fuer Kerntechnik - A.J. Huettl

ATOMWIRTSCHAFT/ATOMTECHNIK 30(1985)11

Stand der Anlage nach Erteilung der Maschinen- und Elektrotechnischen TEG - B.J. Baumgartl u.a.

Kernenergie in der Elektrizitaetswirtschaft in der BRDeutschland 1984 - H.W. Schiffer

Verzeichnis der Kernkraftwerke der Welt - M. Kempken

Arbeitsausnutzung der Kernkraftwerke der Welt im 1. Halbjahr 1985 - A. Szeless, F. Oszuszky

T a g u n g s b e r i c h t e:

Einfluss von EDV-Fortschritten auf die Safeguards-Praxis (Bericht ueber das 7. ESARDA-Jahressymposium in Luettich/Belgien)

Sicherheit von Hochtemperaturreaktoren (Bericht ueber die KTG-Fachtagung in Juelich)

EUROPEAN JOURNAL OF NUCL. MED. 10(1985)1-2

Gradient effects in extravascular water determination using O-15-labelled water under steady state conditions. Theory and error sensitivity - G.-J. Meyer et al.

EUROPEAN JOURNAL OF NUCL. MED. 10(1985)3-4

Nuclear medicine liver function tests for pregnant women and children. 1. Breath tests with C-14-methacetin and C-13-methacetin - P. Krumbiegel et al.

EUROPEAN JOURNAL OF NUCL. MED. 10(1985)9-10

The O-15 continous-inhalation method. Correction for intravascular signal using C-O-15 - P. Pantano et al.

Localisation of C-11-radiopharmaceuticals in the Greene melanoma of hamsters - J.H. Turner et al.

FIZIOLOGICHESKII ZHURNAL (UdSSR) 31(1985)3

Metabolism of H-3-hydrocortisone in liver, kidney and spleen cells of Guinea pigs - A.G. Michenko et al.

An autoradiographic study of the lymphocytotropic function of low-molecular humoral thymus factor LSS in thymectomized thym-splenectomized mice - E. Yu. Giryavenko et al.

J. OF CEREBRAL BLOOD FLOW AND METABOLISM 5(1985)2

Glucose metabolic rate kinetic model parameter determination in humans : The lumped constants and rate constants for (F-18) Fluorodeoxyglucose and (C-11)Deoxyglucose - M. Reivich et al.

Double tracer study of the fine regional blood-brain glucose transfer in the rat by computer-assisted autoradiography - A. Gjedde et al..

Effect of iodoacetate on local cerebral blood flow and glucose metabolism in Cats : A double-radionuclide autoradiographic study - K. Tanaka et al.

J. CLIN. PHARMACOLOGY 25(1985)4

Staggered stable isotope administration technique for study of drug distribution - J.E. Evans et al.

MEDIZIN UND SPORT 25(1985)1

Zur Ermittlung von Parametern des Proteinumsatzes mit Hilfe der N-15-Tracertechnik bei Sportlern - P. Krumbiegel et al.

DRUG METABOLISM REV. 15(1985)7

Determination of toxic pathways of metabolism by deuterium substitution - L.R. Pohl u. J.R. Gillette

UKRAINSKII BIOCHIMICHESKII ZHURNAL 57(1985)3

Chemiluminescent determination of copper in blood serum - V.A. Baraboi et al.

Chemiluminescent determination of iron, copper, and cobalt in blood serum - O.A. Zaporozhets et al.

Effects of coenzymes of pyruvate dehydrogenase

and mitochondria protein on accumulation of S-35-Ripoic acid in them - A.Y. Rozanov et al.

J. PHARMACEUTICAL SCIENCES 74(1985)5
Stable-isotope methodology for the bioavailability study of phenytoin during multiple-dosing regimes - Y. Kasuya et al.

Radiolabeling of intact tablets by neutron activation for in vivo scintigraphic studies - A. Parr et al.

KHIMIKO-FARMATSEUTICHESKII ZHURNAL (UdSSR) 19(1985)4
Synthesis and pharmacokinetics of tritium-labelled thiacoccid - A.D. Shestakov et al.

BIULLETEN EKSPERIMENTALNOI BIOLOGII I MEDITSINY 98(1984)10 (UdSSR)
Effect of naloxone and thyrotropin releasing hormone on respiration in acute hypoxia - V.A. Voinoc et al.

NATURE 315(1985)6019
Terrestrial Xe isotope constraints on the early history of the Earth - M. Ozimo et al.

Reconstruction of prehistoric plant production and cooking practices by a new isotopic method - C.A. Hastorf, H.J. De Niro

PHYS. REV. B 31(1985)10
Anomalous isotope effect for hindered rotations : H_2O adsorbed on metal surfaces - J.E. Mueller

J. AMER. CHEM. SOC. 107(1985)11
Heavy-atom kinetic isotope effects in the acid-catalyzed and thermal rearrangements of 2,2'-hydrazonaphthalene. Transition-state differences in their concerted rearrangements - H.J. Shine et al.

Z. FUER CHEMIE 25(1985)3
Determination of traces of Os with dithizone by substoichismetric isotope dilution analysis - G. Robisch et al.

APPLIED PHYSICS B 37(1985)2
Laser isotope separation of T using CF_3CT BrF/ CF_3CHBrF - K. Takeuchi et al.

ZH. FIZICHESKOI KHIMI 59(1985)4 (in Russian)
Investigation of isotopic effects at the low temperature ethanol photolysis - A.A. Obynochnyi

QUART. J. ROYAL METEOROL. SOC. 111(1985)468
Isotopic study of hail : the $\delta D-\delta^{18}O$ Relationship and the growth history of large hailstones - J. Jouzel et al.

ANALYT. CHEM. 57(1985)7
Multisample conversion of water to hydrogen by Zn for stable isotope determination - C. Kendall et al.

Pyrolytic decorboxylation of aromatic acids as a means of isotopic analysis - J. Santrock et al.

Isotopic analysis based on the mass spectrum of CO_2 - J. Santrock et al.

J. CHEM. SOC., CHEM. COMM. (1985)11
Isotope labelling studies on the Ru-catalyzed hydroformylation of ethylene : indirect evidence for catalysis at intact clusters - G. Suess-Fink et al.

CONTRIBUTIONS MINEROLOGY PETROLOGY 89(1985)2-4
Precambrian tholeiitic-dacitic rock-suites and Camprian ultramafic rocks in the Pennine nappe system of the Alps : Evidence from Sm-Nd isotopes and rare earth elements - P. Stille et al.

The Criffell zoned pluton : correlated behaviour of rare earth element abundances with isotopic systems - W.E. Stephens et al.

Preliminary investigations of 18-O/16-O and D/H compositions in rhyo-ignimbrites in the Inhihaou (in Zize) magmatic center. Central Ahaggar, Algeria - S. Fourcade, M.Javoy

NUCL. INSTRUM. METHODS A 236(1985)1
An electron-beam-generated-plasma ion sources for on-line isotope seperation - J.M. Nitschke

J. APPLIED PHYSICS 57(1985)11
Parametric analysis of isotope enrichment in a vacuum-arc centrifuge - M. Kristnan, R.P. Prasad

J. CHEM. PHYSICS 82(1985)11
Isotope effects on the population of metastable H_3 and D_3 molecules in neutralized ion beams - Slung-Joon Jeon et al.

Quantum mechanical theory of isotope effect on thermally activated hydrogen migration on W(110) - K.F. Freed

J. MOLECUL. SPECTROSCOPY 111(1985)2
IR spectrum of isotopic CO_2 in the region of bending fundamental $\nu 2$ - K. Jolma

Analysis of ν_2 of $H_2{}^{33}S$ and $H_2{}^{34}S$ - W.C. Laue et al.

The ν_2 band of $^{15}NH_3$ - R.D'Cunha et al.

The pentad rotation - vibrational states of $^{12}C\ D_4$ - Jens-Erik Lolek et al.

LA RECHERCHE (1985) 162 (in French)
La separation isotopique par laser - W. de Ruiter

Fixation de l'azote et agriculture tropicale - Y. Dommergues et al.

LA RECHERCHE (1985) 163 (in French)
Le quarkonium - A. Martin et al.

LA RECHERCHE (1985) 164 (in French)
La chimie interstellaire - J. Lequeux

ANGEW. CHEMIE 97(1985)3
Monoklonale Antikoerper : Chemie, Funktion und Anwendungsmoeglichkeiten - F.R. Seiler et al.

Hexafluorpropenoxid - eine Schluesselverbindung der organischen Fluorchemie - H. Millauer et al.

"In-out"-isomere Grosshohlraeume und ihre unterschiedlichen Gastselektivitaet - J. Franke et al.

Carbocyclische Grosshohlraeume zum selektiven Einschluss organischer Gastmolekuele in waessriger Loesung - F. Voegtle et al.

INFRARED PHYSICS 25(1985)3
Optically - pumped pulsed laser action in isotopically - substituted DCN - H.N. Rutt et al.

KINETICS AND CATALYSIS 25(1984)5, Part 2
Evaluation of the role of the proton return step in the kinetics of the isotopic exchange of hydrogen in cumene in a potassium dimethylsulfoxide catalytic system - J.O. Chapiro et al.

MARINE CHEMISTRY 16(1985)3
The isotopic mass balance of sulphur in oceanic sediments (The Pacific Ocean ar an example) - A.Yu. Lein

of separation and purification of Na^{99m}TcO$_4$ mainly because of the availability of low specific activity ^{99}Mo within the country. The indigenously developed technology has been satisfactorily used by us at the Radiation Medicine Centre (the largest single user in the country) for the past decade with good results. The shortcomings present in the earlier version of the method [6] — arising mainly from the radiation dose have been largely overcome by realigning the various assemblies or using appropriate shielded or remote-handled devices or by semi-automating many of the processes. The method has resulted in good yields (75—88%) and a product of good radiochemical and radionuclidic purity; and has been operated with radiation exposures far below the maximum permissible levels for occupational workers. (A detailed account of the purity aspects as well as radiation exposures encountered in these operations would be dealt with in a separate communication). More than 75·0 Ci amounts of Na^{99m}TcO$_4$ were solvent extracted and converted into the medicinal form using this technology in the year 1983. A variety of ^{99m}Tc-radiopharmaceuticals, viz. bone agents (^{99m}Tc-MDP/-HEDP/-Pyrophosphate/-EDTMP), hepatobiliary agents (^{99m}Tc-DIPIDA/-p-butyl-IDA/-dimethyl-IDA), renal agents (^{99m}Tc-DTPA/Glucoheptonate), reticulo endothelial agents (^{99}Tc-Phytate/-sulfur colloid/-Sb$_2$S$_3$ colloid), lung agents (^{99m}Tc-HSAmicropheres/-MAA), ^{99m}Tc-labelled (in vivo) RBC for blood pool etc. have been prepared using the solvent extracted and purified Na^{99m}TcO$_4$, and have been satisfactorily used in patients.

Earlier workers have designed different (manual to the fully automated) types of solvent extraction systems to prepare Na^{99m}TcO$_4$ [3, 4, 6 12]. A fully automated liquid-liquid extractor for the preparation of injectable Na^{99m}TcO$_4$ was available from commercial sources in the early 1970 [13, 14]. The Australians and French have also relied upon large-scale automated systems for a number of years to prepare Na^{99m}TcO$_4$ injectable in a central radiopharmacy [4, 15, 16]. The main advantage of a fully automated system is that large amounts of activity levels could be handled and the multiple operations can be performed with little radiation dose and exposures to personnel. Nowadays fully automated S.E. assemblies are not easily available on a commercial basis, and it requires a lot of expertise to fabricate a fool-proof system tailored to customer' s (large or small) requirements. Besides there is a declining trend in the use of such systems. For example, the S.E. process is rarely used in big or small nuclear medicine centres in developed countries. The simplicity of the automated elution procedure (from column chromatographic devices based on fission-produced ^{99}Mo) for routinely preparing Na^{99m}TcO$_4$ has rendered the S.E. procedure more or less obselete. The former requires less expertise to operate the system at the user end as compared with the latter. And also ^{99m}Tc is available in good yield and purity 'on tap' at a fraction of the time required for preparing Na^{99m}TcO$_4$ by the (manual) S.E. process. Despite these obvious advantage we have adopted the manually operated S.E. process for preparing Na^{99m}TcO$_4$ because of limited choice.

A few advantages of the manually operated S.E. process described herein are: (I) it could be used for separating $^{99m}TcO_4^-$ from $^{99}MoO_4^{2-}$ of different (high or low) radioactive concentrations and specific activities, (II) it could be used to extract large or small amounts of activity depending on the individual user's requirements, (III) the process could be (manually) checked/controlled to result in high yields and good quality Na^{99m}TcO$_4$ — by following stipulated guide lines, ensuring MEK is completely removed, avoiding overheating, etc., and (IV) high radioactive concentrations of Na^{99m}TcO$_4$ could be obtained, if required, for good bolus injections. The disadvantages of the method are (I) the time taken and the number of operations required to be carried out for the preparation of the injectable, (II) the likelihood of radiation exposures to personnel performing these operations (unless proper precautions are taken in carrying out each of the operations), and (III) the expertise (a knowledge of radiochemistry, radiopharmacy, radiation protection, health physics) required to effeciently carry out the multiple operations. We have employed the method during the past decade for beneficially separating and purifying Na^{99m}TcO$_4$ on a daily basis from a low specific activity ^{99}Mo. The specific activities ranged from 9·8—300·0 mCi/g of Mo on the first day of extraction. Volumes upto ~ 110 ml have been handled by this technique. The entire operation for the preparation of the injectable Na^{99m}TcO$_4$ and formulation of different dosage forms, i.e. other ^{99m}Tc-radiopharmaceuticals, takes about 45 min. The S.E. process, however, requires attention to detail, and a certain amount of dexterity and skill in manipulating the various operations in order to obtain high yields of Na^{99m}TcO$_4$ of good (radionuclidic, radiochemical, chemical, pharmaceutical) purity. In addition motivation of personnel is called for if the radiation exposures are to be kept within the dose limitation guidelines (ALARA).

Received April 19, 1984
Accepted in revised form July 18, 1985

References

[1] *M. Waldrop*, Chem. Engg. News 57-21 (1979) 16
[2] *P. Richards, W. D. Tucker, S. C. Srivastava*, Int. J. Appl. Radiat. Isotopes **33** (1982) 793
[3] *L. G. Colombetti*, Performance of ^{99m}Tc generating systems, ed. *B. A. Rhodes*, Quality Control in Nuclear Medicine, St. Louis: C.V. Mosby Co. 1977, p. 183
[4] *R. E. Boyd* Int. J. Appl. Radiat. Isotopes **33** (1982) 801
[5] *J. V. Evans, P. W. Moore, M. E. Shying, J. M. Sodeav*, A new generator for technetium-99m, ed. *C. Raynaud*, Nuclear Medicine and Biology, Vol. II — Proc. Third World Congress Nucl. Med. Biol., Paris: Pergamon Press 1982, p. 1592
[6] *R. S. Mani, D. V. S. Narasimhan*, Development of kits for short-lived generator-produced radioisotopes. Radiopharmaceuticals and Labelled Compounds, Vol. I. Vienna: I.A.E.A. 1973, p. 135
[7] *M. Anwar, K. Lathrop, D. Rosskely, P. V. Harper*, J. Nucl. Med. **9** (1968) 298
[8] *M. C. Crews, B. L. Westerman, J. L. Quinn* III, J. Nucl. Med. **11** (1970) 386
[9] *K. A. Lathrop*, Preparation and control of ^{99m}Tc-radiopharmaceuticals. Raidopharmaceuticals from generator-produced radionuclides Vienna: I.A.E.A. 1971, p. 39
[10] *R. J. Baker*, Int. J. Appl. Radiat. Isotopes **22** (1971) 483
[11] *G. D. Robinson*, J. Nucl. Med. **12** (1971) 459
[12] *V. J. Molinski*, Int. J. Appl. Radiat. Isotopes **33** (1982) 811
[13] *D. N. Torren, M. R. Powell*, J. Nucl. Med. **11** (1970) 368
[14] Imaj International Inc. (Nucl. Med. Div. — Allergen Pharmaceuticals), Mektec 99™ — Automated liquid extractor for the preparation of technetium-99m, Brochure, Irvine, Calif., U.S.A., 1974
[15] *R. E. Boyd*, Recent developments in generators of ^{99m}Tc. Radiopharmaceuticals and Labelled Compounds, Vol. I, Vienna: I.A.E.A. 1973, p. 3
[16] *R. Constant, C. Fallais, R. Charlier, P. Bievelez*, Le generateur ^{99m}Tc controle du purete et controle de production, ibid. p. 27

Isotopenpraxis 22 (1986) 2, S. 58 bis 62

Solvent Extraction Technology of ^{99}Mo$-$^{99m}Tc Generator System 2. Radiation Dose and Personnel Exposure Considerations

O. P. D. Noronha

(Radiation Medicine Centre, Bhabha Atomic Research Centre, c/o. Tata Memorial Hospital Annexe, Parel, Bombay)[1])

The chest, wrist and finger doses experienced over a period of ten years by a small group of radiopharmacists (working in one of the largest nuclear medicine centres in India) have been evaluated. These personnel were involved in the preparation of injectable Na^{99m}TcO$_4$ (and formulation of a few ^{99m}Tc-radiopharmaceuticals) by an alternate method that is seldom employed in most countries, viz. the (manually operated) solvent extraction technique. They received fairly high doses in the earlier years. Their radiation burdens were systematically decreased over a period of time by a significant extent by adopting proper radiation management techniques-work procedures, attitudes and by employing a number of radiation safety/protection devices. The ALARA dose limitation philosophy was the motivating force behind this self-interest programme.

Die über einen Zeitraum von zehn Jahren an einer kleinen Gruppe von Radiopharmazeuten (die in einem der größten kernmedizinischen Zentren Indiens arbeiten) beobachteten Brust-, Handgelenk- und Fingerdosen wurden ausgewertet. Dieses Personal befaßte sich mit der Herstellung von injizierbarem Na^{99m}TcO$_4$ (und mit der Formulierung einiger weniger ^{99m}Tc-Radiopharmaka) nach einer Methode, die in den meisten Ländern selten angewandt wird, nämlich der (von Hand durchgeführten) Lösungsmittelextraktion. In den ersten Jahren waren sie recht hohen Dosen ausgesetzt. Ihre Strahlungsbelastung wurde systematisch verringert durch Einführung entsprechender Strahlensicherheitsmaßnahmen und durch den Einsatz einer Reihe von Strahlenschutzeinrichtungen. Die motivierende Kraft, die hinter diesem Strahlenschutzprogramm stand, war die ALARA-Dosis-Begrenzungs-Philosophie.

Keywords

chest dose; finger dose; molybdenum 99; nuclear medicine; pertechnetates; radioisotope generators; radiopharmaceuticals; solvent extraction; technetium 99m; wrist dose

Introduction

Unlike in most countries of the world we in India have adopted the solvent extraction method for the separation (from Na$_2$^{99}MoO$_4$) and purification of Na^{99m}TcO$_4$ required in clinical nuclear medicine. The purified product is further processed to yield a pharmaceutical dosage form of Na^{99}TcO$_4$ in physiological saline suitable for parenteral use [1, 2]. The process generally employed in India involves the manual handling of large amounts of ^{99}Mo and ^{99m}Tc radionuclides in the solution form (as unsealed or open radioactive sources) during the several radiochemical and radiopharmacy procedures [2]. Often times these sources are handled in an adequately exhausted fume cupboard in the respective hospital radiopharmacy. The manual handling of large amounts of both these radionuclides during processing within the precincts of a (sometimes inadequately equipped) hospital radiopharmacy could pose a potential radiation risk and hazard to the operating personnel, unless adequate prevention/protection measures are implemented to safeguard against radiation exposures. This report summarises our decade — long experience in containing the radiation field emanating from these radioactive sources during their handling, as well as in evaluating the personnel doses received during the operation of these solvent extraction based radiophrmacy procedures.

Materials and Methods

An indigenously fabricated extraction assembly was used for the separation of Na^{99m}TcO$_4$ from Na$_2$^{99m}MoO$_4$ [3]. This allglass unit was externally shielded with 2·5 cm thick lead. The (^{99}Mo + ^{99m}Tc) sources ($\sim$ 200 mCi) were consigned in 10 or 15 ml capacity glass vials in adequately shielded (2·5 – 3·0 cm) lead pots. The ^{99}Mo source depleted of ^{99m}Tc (after the removal of ^{99m}Tc by solvent extraction) was contained and stored in an extra thick lead pot for subsequent use on the following day(s). This source was kept at the rear of the extractor in a fixed place and was not moved around (even remotely) nor manually handled. (In the earlier years the shielded but open source was moved around to collect the ^{99}Mo solution and sometimes remotely handled.) The several radiochemical operations were performed in an adequately shielded (0·6 cm lead sheet on all sides, and 5·0 cm thick interlocking lead bricks to a height of $\sim$ 30 cm in front) and exhausted fume cupboard. The radiopharmacy operations were carried out in a clean enclosure adjacent to the fume cupboard to ward off any cross contamination [1, 2]. Normal radiochemical safety procedures like wearing laboratory overcoats, latex rubber surgical gloves, etc. were generaly followed during these operations. In the early days (till mid 1977) many of the above operations were carried out manually with the use of a bare minimum of lead shielding (only the lead wall in front of fume cupboard) and remotely-operated gadgets like $\sim$ 20 – 25 cm long forceps/pair of tongs, etc. It was

[1]) Address: Bombay 400012, India

also carried out in a smaller and restricted work-area. During that period only about $100-150$ mCi amounts of ^{99}Mo + ^{99m}Tc were handled per day. Subsequent to this the activity was augmented to $\sim 300-475$ mCi per day. Gradually over the years and with gain of experience many more improvised lead-shielded gadgets/assemblies and remote-handling devices were introduced to safeguard the radiopharmacy personnel against radiation. Some of these radiation safety/protection measures followed were: the use of extra-long ($\sim$ 35 cm) decapping tool to deseal the ^{99}Mo vial, various types of extra-long ($>$ 50 cm) pair of forceps/tongs, various types of lead-shielded assemblies for: the stored ^{99}Mo source in a 100 ml vial, beakers/*Erlenmeyer* flask containing ^{99m}Tc solutions, the hot plate during evaporation of MEK solvent, and for the vial containing stock/dispensed solutions of ^{99m}Tc. Lead-lined apparel like aprons (0·5 mm Pb equivalent) and gloves (0·16 mm) (since mid 1981 and beginning 1983 respectively), table-top shields of lead and leaded glass, thick leaded glass windows, and syringe shields and Radiglasses (occasionally) were also used. Semi-automated devices like remotely (electrically) operated and controlled low vacuum cum pressure pump, etc. [1, 2] were also used in the later years. On most occasions the preparation of $Na^{99}TcO_4$ injectable was performed by 2 – 4 operators working in rotation during the period of a week.

Chest and wrist film or $CaSO_4$ (Dy) TLD badges (the latter since 1981) were generally worn on the chest and wrist by the personnel. However, with the introduction of the lead-lines aprons they were generally worn on the rear side of the lead apron. These badges were processed and doses evaluated fortnightly/monthly by the Personnel Monitoring Section of the Division of Radiation Protection, B.A.R.C. Finger dosimetry was carried out by attaching thin 1 cm^2 TLD (LiF) discs onto the distal phalanges of the fingers during these solvent extraction based operations [2]. These studies were carried out with or without the use of lead-lined gloves.

Results and Discussion

The solvent extraction technology pertaining to the preparation of $Na^{99m}TcO_4$ injectable essentially involves the carrying out several radiochemical and radiopharmacy processes in a sequential manner. The manual handling of these operations could result in a radiation dose. The main radiochemical operations were the (assay of ^{99}Mo +^{99m}Tc source vial in a dose calibrator, decapping the source vial) charging the source into the extractor, mixing of the source with an organic solvent (MEK), removal of ^{99}Mo depleted of ^{99m}Tc, purification of ^{99m}Tc by scrubbing, further purification of ^{99m}Tc by passing the organic phase through a column of alumina, stripping off the organic solvent, and dissolving $Na^{99m}TcO_4$ in physiological saline. The important radiopharmacy operations were steam sterilisation (in the early days)/membrane (0·22 μm) filtration, assay of ^{99m}Tc stock solution, dispensing of $Na^{99m}TcO_4$, formulation of $2-4$ different ^{99m}Tc-radiopharmaceuticals and reassay of these formulated ^{99m}Tc agents. These operations were performed only once a day.

The weekly whole body (chest) dose equivalent in mrems received by select personnel operating the solvent extraction based radiopharmacy schedules is given in Tab. 1. The data generally relate to $\sim$ 2 schedules per operator per week. In some cases it could represent one but rarely three operations per week. It is evident that the dose experienced is operator dependent. It can be seen that over the years there has been a steady decrease in the weekly whole body dose equivalent received by the personnel. This is so despite a two-or threefold increase in the activity handled per day in the hospital radiopharmacy subsequent to mid 1977. The annual ^{99m}Tc activity solvent extracted and converted into an injectable form for clinical use is also indicated in Tab. 1. This showed a fourfold increase during the period surveyed, i.e. from 1974 to 1983. In the early days the average whole body dose equivalent per person per week was $\sim$ 15 mrem. This corresponded with the handling of $\sim 100-150$ mCi levels of ^{99}Mo$-^{99m}$Tc per day of the week. This was reduced to $\sim$ 3·2 mrem per week (mean of A, B and D), despite augmentation of the activity ($\sim 300-475$ mCi amounts per day). The dose equivalent received per person per Ci of ^{99m}Tc handled in these operations worked out to $<$ 10 mrem and in some cases $\sim 2-3$ mrem (assuming ^{99}Mo$-^{99m}$Tc radiopharmacy schedules are handled twice a week). The

Tab. 1. Weekly whole body dose equivalent (mrem) received over the years by select personnel operating solvent extraction based radiopharmacy schedules

Operator	calendar year										
	1973	1974	1975	1976	1977*)	1978	1979	1980	1981	1982	1983
A	21·3	20·1	15·3	9·6	–	4·4	3·4	6·7	4·7	3·7	2·5
B	10·5	3·6	4·8	5·3	0·8	4·2	3·0	1·9	2·3	0·5	1·0
C				25·0	14·7	7·9	9·7	–	6·4	4·1	–
D					20·2	22·1	–	6·2	7·2	6·6	6·0
E						13·1	8·0	6·5			
F								15·6	10·1	12·5	
G										10·6	10·0
means ± s.d.	15.9 ± 7·6	11·8 ± 11·6	10·0 ± 7·4	13·3 ± 10·3	11·9 ± 10·0	10·3 ± 7·5	6.0 ± 3·3	7·4 ± 5·0	6·1 ± 2·9	6·3 ± 4·5	4·9 ± 4·0
average ^{99}Mo + ^{99m}Tc handled per week, mCi		385	400	775	625	1275	1450	1300	1500	1650	1700
total ^{99}Tc handled per year, Ci		17·1	18·1	34·2	28·0	56·5	64·8	58·1	67·2	73·0	75·6

*) In 1977 the radiopharmacy operations were suspended for sometime and then moved to a more commodious lab. The lab. was gradually fitted with better shielding and other remote-controlled equipment. The (^{99}Mo + ^{99m}Tc) activity was also augmented during the year. Operator D generally wore film badge on the outside of lead apron. Operators E, F, G were relatively new recruits.

dose equivalent received by skilled novices, i.e. personnel knowledgeable in radiochemical and radiopharmacy principles/procedures but new to radiation protection practices was ~ 10–16 mrem/week (data not included in Tab. 1) while handling ~ 400 mCi amounts of $^{99}Mo-^{99m}Tc$ radionuclides per day. This was reduced to a considerable extent (~ 4–7 mrem) with experience, as can be seen from a survey of the data in Tab. 1. Also, over the years the mean dose equivalent experienced by the radiopharmacy personnel showed a gradual decline from ~ 15·0 to ~ 5·0 mrem/week.

A similar pattern of dose equivalent was observed by using a pen-type pocket dosimeter attached to the lapel of the shirt (outside the lead apron) of one of the personnel. The dose registered was ~ 50 mrem over a period of 9 weeks. This involved the separation of 6026 mCi amounts of ^{99m}Tc on 22 different occasions, i.e. an average activity of 274 mCi per occasion. The dose equivalent registered was ~ 8·0 mrem/Ci of ^{99m}Tc extracted, or ~ 2·3 mrem/extraction, or ~ 5·5 mrem/week (assuming 2·5 extractions/week).

The weekly wrist dose equivalent received over the years by select radiopharmacy personnel operating solvent extraction schedules is shown in Tab. 2. The dose equivalent was high in the early years, ~ 90·0 mrem/week, in 1973. This was decreased to ~ 13 mrem/week over a ten-year period with the increased use of radiation safety/protection devices. It can be seen that despite the decrease the process could result in high dose equivalents to personnel operating these processes in a complacent or casual manner. The dose equivalents received by novices handling the solvent extraction schedules was of the order of 30–60 mrem/week.

Tab. 2. Weekly wrist dose equivalent (mrem) received over the years by select personnel operating solvent extraction based radiopharmacy schedules

operator	year										
	1973	1974	1975	1976	1977	1978	1979	1980	1981	1982	1983
A	94·1	59·8	43·1	45·7	–	7·0	8·9	19·6	18·0	28·8	8·0
C				28·0	48·6	72·1	40·2	–	23·9	46·5	–
D					26·3	35·1	–	19·7	15·8	21·4	18·0
E						62·1	43·9	40·0			
F									41·9	98·0	
G										28·7	14·0

Only latex rubber gloves were worn till 1982. Lead-lined gloves were sometimes worn in 1983.

The finger dose received by select personnel operating these types of radiopharmacy procedures is given in Tab. 3. The exposures received were measured with TLD dosimeters attached to the fingers of personnel (wearing only latex rubber surgical gloves). It can be seen that the finger dose received varied over a wide range of values, and was not evenly distributed among the fingers. The thumb, index and middle finger of the right hand experienced a considerable amount (~ 100 mR) of dose per operation while handling 400–500 mCi amounts of $^{99}Mo-^{99m}Tc$. However, the actual close received was again found to be operator dependent. The hypothetical data included in Tab. 3 (corresponding to the mean of doses received by skilled operators A and B) indicates the likely doses that could be experienced by the fingers of skilled personnel handling 450 mCi amounts of $^{99}Mo + ^{99m}Tc$. The likely levels for handling 100 mCi amounts of these nuclides is also indicated in Tab. 3. The dose could be of the order of 10–20 mR per schedule. In actual practice some of the operators received two or three times the mean exposure levels. One of the operator (K), however, received lesser dose. (The data in this particular case are not truly representative since this operator did not handle fresh stock solutions of $^{99}Mo + ^{99m}Tc$. The particular operation was carried out towards the expiry of ^{99}Mo-generally used for 7 days). It was also observed that accidental contamination of the hand-gloves with a speck of the 'hot' solution could result in a manyfold increase in the finger dose (operator J). As compared to the above the dose received by the finger-tips protected with lead-lined gloves was of the order of 10–30 mR during the handling of 925 mCi amounts of $^{99}Mo + ^{99m}Tc$ and the processing of 716 mCi amounts of ^{99m}Tc obtained therefrom. These latter operations were performed on three different days (extracted once a day). This amounted to a fingertip dose of only 3–10 mR per operation schedule of ~ 310 mCi ($^{99}Mo + ^{99m}Tc$) and processing ~ 240 mCi amounts of ^{99m}Tc obtained therefrom.

The doses registered on the chest, wrist and fingers are indicative of the radiation risk involved in the technology used as well as the responsiveness/awareness of the operators concerned to implement radiation safety/protection practices during the several operations. The mean annual chest dose equivalent per operator performing these solvent extraction based radiopharmacy schedules amounted to ~ 800 mrem (in 1973) and ~ 250 mrem (in 1983) for a 50 working-week year. The corresponding wrist dose worked out ~ 4500 mrem (in 1973) and 650 mrem (in 1983). The chest and especially the wrist dose were high (a fifth and a sixteenth of maximum permissible levels respectively) in the earlier years. They were decreased by three and seven fold respectively over the past few years. This was achieved by deliberately employing better work techniques including radiation safety/protection devices

Tab. 3. Finger dose (mR) received by select personnel operating solvent extraction based radiopharmacy schedules

Operator	^{99}Mo-^{99m}Tc	right hand finger dose					left hand finger dose				
	activity handled mCi	thumb	index	middle	ring	little	thumb	index	middle	ring	little
G	510	119	116	86	65	34	102	58	88	54	162
H	398	70	82	90	38	35	87	82	88	62	128
mean of G & H	450	95	99	88	52	35	95	70	88	58	145
(normalized values)	100	21	22	19	11	8	21	15	19	13	32
I	490	181	193	125	84	62	403	306	179	127	97
J	420	242	228	207	73	38	128	258	742	138	42
K	200	16	38	23	–	–	24	19	25	–	–

(mainly lead-lined gloves in the case of wrist and finger dose). The doses to the fingertips must have been exceedingly high in the earlier years when only latex rubber surgical gloves were worn by the operators. This is evident from the high dose regimens registered. (The wrist dose is an indicator of the hand and finger dose — the dose to the fingertips being about seven times the dose registered at the base of the fingers [4]. The wrist dose is also indicative of the work practices actually employed by each of the operators, including the ingenuity/expertise practised to implement radiation safety procedures.

The whole body dose received by the personnel was small especially during the later years and well within the maximum permissible limits as currently understood [5]. Similarly, the dose experienced by the unprotected fingers (without lead-lined gloves) was also within the maximum permissible limits. However, the finger dose received by some of the personnel was on high side to warrant concern especially their long-term exposures, should they be repeatedly or daily involved in radiopharmacy operations. The dose received dependend very much on individual work habits/practices. One of the contributing factors for the high dose levels in certain individuals was insufficient motivation to scruplously avoid 'hot' sources and to adopt correct (radiation-safe) work practices. Their primary motive was to succesfully complete the 'more important' radiochemical and radiopharmacy operations within a reasonable time period (in order to adhere to appointed patient schedules). This resulted in a general apathy towards 'less important' radiation safety/protection work practices, especially those requiring attention to minor details. It was also observed that operators working under stress or 'rush' situations (e.g. when prevailed upon to complete the task at the earliest, especially if the operations were started late) generally tend to use radiation unsafe short-cut methods. These could result in high levels of exposures. Also, heavy contamination of apparel could result in radiation exposures to localised regions of the body. The lead-lined hand gloves provided a good measure of radiation protection to a part of the arm, wrist, hands and fingertips.

One of the constraints of the manual (solvent extraction) method of preparing $Na^{99m}TcO_4$ is the manual requirement of handling large amounts of both ^{99}Mo and ^{99m}Tc within the precincts of (a sometimes less well-equipped) hospital radiopharmacy. The operation invariably results in a radiation field unless it is fully contained and safeguarded. The latter requirement could be achieved by using appropriate shielding and automation (e.g. remotely operated master slave manipulators with solenoid valve (control devices, or sequentially controlled by means of a computer). Since the latter does not apply for manually operated systems there could be a possibility that personnel operating these processes may be exposed to at least some levels of the radiation field and consequently receive personnel exposures. (The use of radiation unsafe work practices would result in higher exposures. Similarly, personnel working in inadequately equipped facilities could, in addition, receive higher exposures). The actual levels of the radiation fields that may be encountered while handling 400—500 mCi amounts of ($^{99}Mo + ^{99m}Tc$) and ^{99m}Tc alone were of a high order of magnitude. For example, the radiation fields (as measured with a β, γ exposure rate meter — a gun monitor) associated with an unshielded $\sim$ 200 mCi source of $^{99}Mo + ^{99m}Tc$ contained in a standard sized 10 ml flint glass vial (U.S.P./type 1) were $>$ 5000 mR/h at close contact and $>$ 1000 mR/h at $\sim$ 20 cm (the standard length of a commonly used appliance such as a pair of tongs) from the source. So also the radiation fields of an unshielded $\sim$ 400 mCi ^{99m}Tc source were of a similar order of magnitude. The corresponding dose rates of even a $\sim$ 275 mCi amount of ^{99m}Tc — an activity commonly encountered in several nuclear medicine clinics — were $\sim$ 4500 and $\sim$ 600 mR/h respectively. The high radiation field emanating from unshielded sources of ^{99}Mo and ^{99m}Tc assumes added significance because of the element of risk (of exposure) involved while repeatedly performing these 'hot' operations; (sometimes daily by the same operator in smaller radiopharmacy laboratories). It is possible that some of these operations may be carried out with bare hands (covered only with surgical gloves) by novices, unless they are strongly motivated to desist from employing unsafe radiation practices.

External personnel exposures are an inevitable occupational hazard of the manually-operated solvent extraction based technology. The radiochemical operations involved in this technology require the initial handling of both the nuclides $^{99}Mo + ^{99m}Tc$, and subsequently of ^{99m}Tc alone. The radiopharmacy operations are restricted to handling only ^{99m}Tc. Because of the intrinsic nature and energy characteristics of ^{99}Mo (β, γ) and ^{99m}Tc (140 keV γ) the radiations emanating from the latter nuclide can be easily contained by appropriate lead shielding (half value layer = 0·03 cm). Therefore the operations involving ^{99m}Tc pose a less radiation risk in terms of dose as compared to those operations involving the handling of both $^{99}Mo + ^{99m}Tc$. Analysis of our personnel exposures during the later years (Tab. 1 and 3) and comparison with published data of exposures received by radiopharmacy personnel involved in eluting, dispensing and formulating of ^{99m}Tc-radiopharmaceuticals (but using $Na^{99m}TcO_4$ obtained from column chromatographic based $^{99}Mo - ^{99m}Tc$ generator systems) [4, 6—13] suggests that personnel exposures received in both cases are comparable to each other. (The column chromatographic procedure precludes the handling of $^{99}Mo + ^{99m}Tc$ as open solution sources at the user end.) This is so despite the many varied type of operations involved in the solvent extraction procedure. Therefore, it could be deduced that most of the exposures are derived while performing radiopharmacy-based operations — mainly dispensing and formulating ^{99m}Tc radiopharmaceuticals, since the elution (of $Na^{99m}TcO_4$) procedures generally adopted with column chromatographic systems are relatively simple and safe. A plausible explanation of this apparent anomaly is that in experienced hands the radiochemical operations connected with solvent extraction system can be carried out safely, i.e. without actually touching the 'hot' sources or coming into close or proximate contact with them. This could be achieved by adequate contact shielding and using remote-controlled procedures as well as an assorted variety of radiation protection gadgets. Such a situation is not easily achieved with the manual type of radiopharmacy operations as described here. The operator was perforced to come into closer contact (at times) with 'hot' ^{99m}Tc solutions. At our centre, however, the sterile (membrane) filtration was routinely carried out using a remote — controlled and shielded assembly [1]. The use of this contrivance decreased the dose to the fingers, wrist and hands by a significant extent. The dispensing and formulation of 2 — 4 other ^{99m}Tc-radiopharmaceuticals were however performed by withdrawing the 'hot' ^{99m}Tc solutions contained in glass vials (placed in adequately shielded — bottom and sides — lead pots) using unshielded disposable syringes. Hence the doses to the fingers. Another unsafe radiation practice that was followed in the earlier period was to allow the ^{99m}Tc source vial to protrude out of the lead pot at the top — for easy accessibility. This practice also could have contributed to the finger dose. Subsequent to the survey of finger dose many more stringent radiation safety procedures were adopted at our Centre in order to reduce personnel exposures.

The high radiation fields that generally result in such a manually operated system can only be contained by employing well established radiation protection measures. They are the use of appropriate lead shielding for containing the 'hot' sources, the use of remote — handled equipment for increasing the distance, and by decreasing the contact time or time spent proximate to these sources, and by (semi) automation. A process operated on these lines would also be expected to reduce personnel exposures. A semi-automated process assembly for preparing Na^{99m}TcO$_4$ by the solvent extraction procedure and incorporating many of these features has recently been described [2]. The other radiation safety/protection measured adopted by us included the frequent rotation of operators engaged in these programmes, the use of appropriately fitting vial shields for fully containing the 'hot' solutions, the use of flexible lead-lined aprons and extra-long hand gloves, the use of lead and leaded glass (window) syringe shields (used at times), lead tinted goggles, etc. (Washing the hand gloves during operations helped to remove adhering specks of contaminants, if any.) In addition, all the radiopharmacy operations were carried out behind lead and leaded glass table-top barriers. The augmentation of these measures resulted in still lower exposure levels (with respect to 1981). Similarly the finger doses were slo also reduced to $\sim 3-10$ mR per operation by using long lead-lined gloves. These measures were adopted in keeping with the recent trends of reducing the radiation dose to ALARA (as low as resonably achievable) levels.

The manual solvent extraction method is a more elaborate and involved technique (as compared with the column chromatographic method) of preparing the injectable Na^{99m}TcO$_4$. It involves the handling of both ^{99}Mo + ^{99m}Tc and ^{99m}Tc as open ('wet') sources during the several radiochemical and radiopharmacy operation unlike in (relatively 'dry') column chromatographic procedures wherein only Na^{99m}TcO$_4$ is eluted out at the user end. Also these operationshave to be performed by skilled personnel trained in radiochemical, radiation protection, associated health physics aspects as well as in radiopharmacy procedures. In practice this technology requires skill, dexterity and attention to detail while manipulating/manoeuvring the several 'hot' processes. In addition, the personnel must be imbued with a sense of radiation safety/protection work attitude in order to positively avoid/ward off unnecessary exposures arising from the radiation fields associated with the 'hot' sources. In experienced hands the entire process can be safely performed even from the radiation point of view.

The manually operated solvent extraction based technology is the sole method adopted in India for obtaining Na^{99m}TcO$_4$ radiopharmaceutical required in diagnostic nuclear medicine. (The Radiation Medicine Centre is the largest single user of Na^{99m}TcO$_4$ in the country.) There is a paucity of data in the literature regarding the radiation dose and exposure levels associated with this technology, as opposed to the data available on the use of column chromatographic ^{99}Mo – ^{99m}Tc generator systems. So far only Iyer and Dhond [14] have surveyed the occupational doses received by nuclear medicine personnel engaged in these types of operations. The present study was performed by a self interest group of radiopharmacists directly involved in these operations, and interested in decreasing personnel exposures to ALARA levels.

Received August 22, 1984

References

[1] *O. P. D. Noronha, A. B. Sewatkar,* Solvent extraction technology of ^{99}Mo – ^{99m}Tc system: Design and operational considerations, Bhabha Atomic Research Centre Rep. B.A.R.C. – 1215 (1983) 125 p.

[2] *O. P. D. Noronha,* Isotopenpraxis **22** (1986) 53

[3] *R. S. Mani, D. V. S. Narasimhan,* Development of kits for short-lived generator-produced radioisotopes. Radiopharmaceuticals and Labelled Compounds, Vol. 1. Vienna: I.A.E.A. 1973, p. 135

[4] *J. E. Burr, R. Berg,* J. Nucl. Med. Tech. **5** (1977) 158

[5] Annals of the ICRP (publication 26), statement of an Recommendations of ICRP (1980 meeting and Addendum), Oxford: Pergamon Press 1981

[6] *M. G. Mayes, R. L. Douglas, M. W. Carter, T. M. Grady, E. van der Smissen,* Study of radiation exposures from technetium generators at three hospitals. Health physics in the Healing arts, Rockville, Md.: U.S. Dept. Health. Educ. Welfare (F.D.A.) 1973, p. 222

[7] *M. H. Lombardi, W. L. Beck, R. J. Cloutier,* Survey of radiopharmaceutical use and safety in sixty-nine hospitals, ibid p. 227

[8] *B. M. Branson, R. G. Hoops, R. J. Grant,* Assessment of personnel dose from ^{129}Cs and ^{99m}Tc in nuclear medicine laboratory and resulting dose reduction techniques, ibid p. 612

[9] *E. A. Harbottle, R. P. Parker, R. Davis,* Brit. J. Radiol. **49** (1976) 612

[10] *B. Ahluwalia, E. W. Allen, G. Basmadjian, R. Ice,* Health Phys. **40** (1981) 728

[11] *G. A. Lis, S. M. Zu'bi, S. M. Brahmavar,* J. Nucl. Med. Tech. **9** (1981) 91

[12] *P. Schurnbrand, H. Schicha, H. Thal, D. Enrich,* Eur. J. Nucl. Med. **7** (1982) 237

[13] *J. Trujillo, S. Krinsky, B. Wilson, G. Young, E. Teague, K. Kamer,* J. Nucl. Med. Tech. **10** (1982) 197

[14] *P. S. Iyer, R. V. Dhond,* Health Phys. **39** (1980) 576

Isotopenpraxis 22 (1986) 2, S. 63 bis 64

Investigation of Mechanically Treated Iron Electrodes by Means of Autoradiography and *Kossel* Technique*)

A. Pinkowski[1]), *W. Hübner*[1]), *H.-J. Ulrich*[2])

([1]) Academy of Sciences of the GDR, Central Institute of Physical Chemistry; [2]) Technical University Dresden, Dep. of Fundamentals of Engineering Construction)

On anodically biased polycrystalline iron electrodes under the influence of an in-situ mechanical treatment in the pre-passive potential range an enhanced, grain-selective reactivity, resulting in an iron-phosphate precipitate was revealed The different reactivity of the respective grains may be explained about their varying surface charge.

An anodisch polarisierten Eisenelektroden wurde unter dem Einfluß einer in-situ erfolgenden mechanischen Bearbeitung im Vorpassivgebiet eine erhöhte, selektive Reaktivität einzelner kristallographischer Ebenen festgestellt, die zur Abscheidung einer Eisenphosphatschicht führte. Als Ursache für die unterschiedliche Reaktivität einzelner Ebenen wird deren unterschiedliche Oberflächenladung diskutiert.

Keywords

autoradiography; electrodes; iron phosphates; phosphorus 32; polycrystals; properties; surface

Under anodic bias of iron in phosphate electrolytes in the transition region between active dissolution and passivation a pre-passivation of iron occurs [1, 2].

It is the aim of the present work to investigate the influence of a mechanical treatment (MT) on the formation of the prepassive layer. For that purpose a rotating disk electrode (rde) of purity 99·97% and of diameter 6 mm was used. It was polarized potentiostatically at −500 mV, i.e., in the pre-passive potential range (all potentials refer to the saturated calomel electrode). The desired MT was realized about immersing the rde in a bed of glass beads of diameter 1 mm [3]. Caused by the revolution of the rde a MT was achieved (cf. Fig. 1).

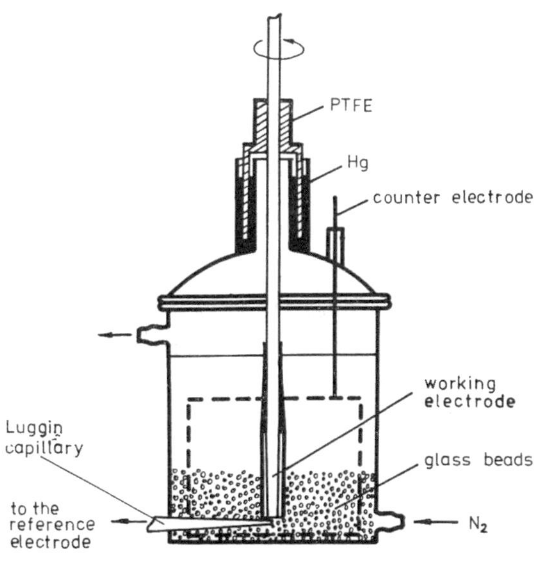

Cell for electrochemical measurements

Fig. 1.

Photographs of the electrode surface after 30 min polarization under MT at −500 mV revealed differences in the layer formation for particular faces of the polycrystalline iron (cf. Fig. 2).

Fig. 2. Photograph of polycrystalline iron sample

For closer inspection of these differences in the layer shape autoradiography [4] and *Kossel* technique were applied.

The electrolyte (1/15 mol · l^{-1} KH_2PO_4 + 1/15 mol · l^{-1} K_2HPO_4 of pH 6·8) was labelled with $KH_2{}^{32}PO_4$. As can be seen from Fig. 3 the counting rate of samples polarized at −500 mV under MT was essentially higher than without MT. Thus the reaction layer formed under MT in the prepassive range consists of any iron phosphate.

Strippingfilm autoradiographs (cf. Fig. 4), however coinciding in shape and structure with the corresponding photographs, showed different phosphate-layer thicknesses

*) The results were represented partly at the XIVth International Symposium on Autoradiography, Reinharsbrunn (GDR), November 19−23, 1984

Addresses: [1]) DDR-1199 Berlin, Rudower Chaussee 5; [2]) DDR 8027 Dresden, Mommsenstraße 13

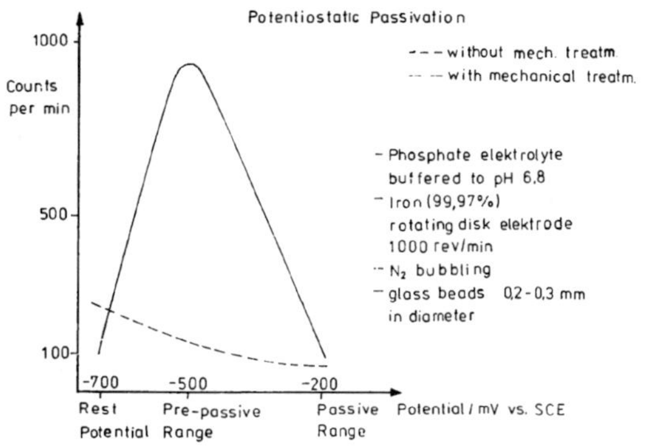

Fig. 3.

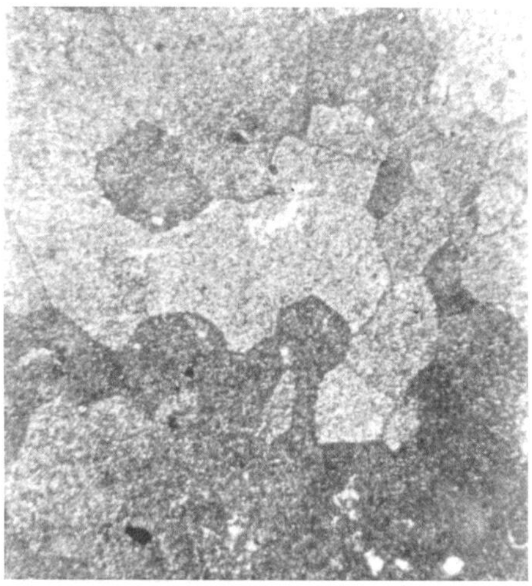

Fig. 4. Autoradiograph of the sample of Fig. 2

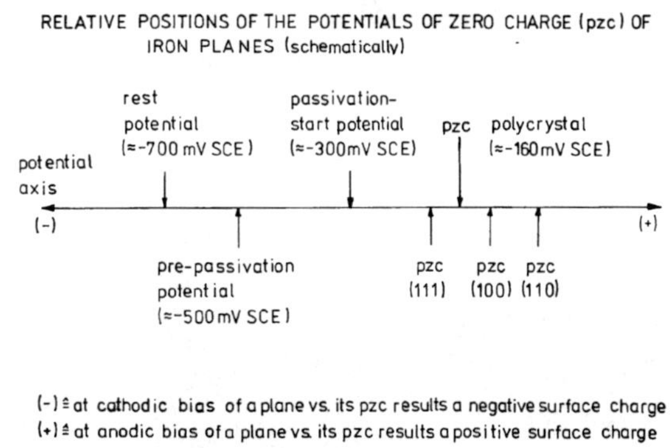

Fig. 5.

on different crystallites. Obviously particular faces of the iron sample show strongly deviating electrochemical reactivity against the phosphate electrolyte. The view of an orientation dependent anisotropy of reactivity of polycrystalline metals is general accepted now.

The crystallographic orientation of selected graines was determined by *Kossel* technique [5]. The comparison of the *Kossel* data with the blackening of the corresponding faces on the autoradiograph gives clear evidence that crystallites near (001) are more reactive than those near (011). Therefore the reactivity decreases with increasing package density of the face and vice versa. This result is in accordance with previous findings on other systems [6]. The interpretation of the connection between electrochemical reactivity and package density shall be based on the concept of the potential-of-zero-charge (pzc) [7].

According to [7] the pzc of metal or semiconductor electrodes determines the adsorption of ions and neutral molecules from the electrolyte as well as the structure of the electrochemical double layer, and therefore the kinetics of charge-transfer reactions across them. The pzc concept plays a key role in the explanation of a lot of experimental results on solid electrodes. Further there exists a strong correlation between pzc, work function, package density and other physical and electrochemical solid-state properties [8]. In general with increasing package density the work function increases and the pzc shifts toward positive potentials. While the pzc of particular faces of a metal differ only by about some 10 to some 100 mV [7], the package density of bcc metals changes (relative to the simple cubic lattice equal to 1) from 1·06 for (110) over 0·75 for (100) toward 0·43 for (111). Unfortunately the pzc of the single-crystal iron faces is not known yet. For polycrystalline iron however the available data are critically reviewed by [7]. Most reliable seems a value equal to −160 mV. While the capacitance of a polycrystalline sample results from summarizing the plane-proportional capacitances of the corresponding faces, the pzc in contrast is close to the pcz of that face with the most negative value, i.e., the pzc of polycrystalline materials is close to the pzc of the face with the lowest package density and the lowest work function [7, 9].

Considering the foregoing the explanation of our results may be given as follows (cf Fig. 5). Evidently the pzc of iron (−160 mV) is positive related to the applied polarization potential (−500 mV), i.e., the surface charge of the iron electrodes at −500 mV has to be negative. The reaction between iron and the phosphate electrolyte requires a transfer of Fe^{2+} ions from the iron lattice toward the electrolyte to form the iron-phosphate precipitate. The negative surface charge may act as a barrier against the transfer of the positive charged iron ions. On faces with the most negative pzc, i.e., with the lowest negative surface charge, the transfer hindering is lower than on faces with more positive pzc, i.e., with higher negative surface charge.

As mentioned above the pzc of the closest packed faces is most positive, it is, closer packed faces should show lower reactivity. Highest reactivity should be found on (111) faces, while faces near (100) should be lie below them. Actually the blackening of the autoradiograph changes in this manner.

Summarizing the anisotropy of the reactivity may be explained about the anisotropy of the pzc, which on their part depend on the package density of particular faces. Thereby the autoradiography allowed an easy determination of the local reactivity.

Received May 22, 1985

References

[1] *A. A. Miligy, O. Geana, W. J. Lorenz,* Electrochim. Acta **20** (1975) 273
[2] *K. Ougura,* J. Electroanal. Chem. **79** (1977) 149
[3] *Leitel, E. A. Pinkowski, K. P. Thiessen,* Z. phys. Chem. (Leipzig) **260** (1979) 448
[4] *A. Pinkowski, K. P. Thiessen,* Nukleonika **21** (1976) 161
[5] *H. J. Ullrich* et al.: Wiss. Z. TU Dresden **28** (1979) 197
[6] *A. Hamelin, J. Lecoeur,* Surf. Sci. **57** (1976) 771
[7] *A. N. Frumkin,* Potentialy nulevogo zarjada (*russ.*) Moskva, Nauka 1979
[8] *L. I. Antropov,* Teoreticeskaya electrochimiya, vyschaya schkola Moskva 1975 (*russ.*)
[9] *R. Parsons,* Surf. Sci. **101** (1980) 316

Isotopenpraxis 22 (1986) 2, S. 65 bis 67

Bestimmung des O/U-Verhältnisses von UO_{2+x} mit röntgenradiometrischen Mitteln

D. Müller, P. Morgenstern

(Zentralinstitut für Isotopen- und Strahlenforschung der AdW der DDR, Leipzig)[1])

Es wird eine radiometrische Methode zur Bestimmung des O/U-Verhältnisses an Kernbrennstoffen beschrieben. Das Analysenverfahren basiert auf der RFA im ultraweichen Röntgengebiet. Als Strahlungsdetektoren wird ein speziell für dieses Energiegebiet (250 eV ... 6 keV) konzipierter PSD eingesetzt. Der erreichbare Meßfehler von ±0,01 O/U-Einheiten ist zwar um ca. eine Größenordnung schlechter als für die bekannten chemischen Methoden. Der Vorteil des hier beschriebenen Verfahrens liegt jedoch in der Schnelligkeit und Zerstörungsfreiheit der Meßmethode.

It is described a method for the determination of the O/U ratio in fuel pellets basing on the X-ray fluorescence in the ultra soft X-ray region. As detection instrument a special proportional scintillation counter is used.The accuracy of the results is nearly 0·01 O/U units. It is actually order of magnitude lower than for the known chemical methods but the advantage of the described procedure is to be an express and nondestructive technique.

Keywords

fuel pellets; gas scintillation detectors; oxygen; uranium; uranium oxides; X-ray fluorescence analysis

1. Einleitung

Unter den in Leistungsreaktoren fast ausschließlich zum Einsatz kommenden keramischen Kernbrennstoffen spielt das Urandioxid wohl eine entscheidende Rolle. Dabei ist in allen Phasen seiner Produktion eine Kenntnis der exakten Stöchiometrie der mit UO_{2+x} bezeichneten Verbindung unabdingar dadurch, daß eine Reihe wichtiger physikalischer sowie technologischer Eigenschaften von dem sogenannten O/U-Verhältnis, d. h. dem Quotienten der Anzahl der Sauerstoff- und Uranatome im UO_{2+x}-,,Molekül", bestimmt werden.

So sorgt eine geringfügige Überstöchiometrie mit Werten $x \approx 0{,}015$ für eine möglichst hohe thermische Leitfähigkeit der Brennelemente und eine geringe Expansion unter Bestrahlungsbedingungen, aber auch für ein großes Rückhaltevermögen von insbesondere gasförmigen Spaltprodukten.

Während des Herstellungsprozesses jedoch gewährleisten Werte von $x = 0{,}03 \ldots 0{,}15$ eine wünschenswert hohe Sinterdichte, was eine anschließende Reduktion des oxidischen Kernbrennstoffes in Wasserstoffatmosphäre erforderlich macht.

Das O/U-Verhältnis erweist sich somit als eine wesentlicher analytischer Parameter für das Ausgangs-, Zwischen- und Endprodukt. Für seine Bestimmung wurden eine Reihe chemischer, physikalisch-chemischer und auch röntgenografischer Methoden erarbeitet, die sich zumeist durch eine hohe erreichbare Genauigkeit auszeichnen. So liefert die O/U-Bestimmung mit Hilfe der Röntgenbeugung, die auf Messungen von Veränderungen der Gitterparameter als Folge des Einbaues von Überschuß-Sauerstoff in Zwischengitter-Positionen basiert, Meßgenauigkeiten unterhalb ±0,001 O/U-Einheiten. Relativ einfach zu handhaben ist die Oxidation des UO_{2+x} zu U_3O_8 bei Ermittlung der Gewichtsdifferenz und erreichbaren Genauigkeiten $\Delta O/U \geqq 0{,}001$, allerdings muß dabei das pulverförmige Urandioxid über mehrere Stunden eritzt werden. Der hohe Zeitaufwand läßt dieses und auch andere elektrochemische und maßanalytische Verfahren (vergleiche etwa [1]) zusammen mit dem zumeist ,,zerstörenden" Charakter als wenig geeignet erscheinen, ein schnelles und unmittelbar steuerndes Eingreifen in den Verarbeitungsprozeß zuzulassen.

Im folgenden soll ein Verfahren zur Bestimmung des O/U-Verhältnisses beschrieben werden, das auf der Anregung von charakteristischer Röntgenstrahlung mit Hilfe von Radionuklidquellen und deren Nachweis mit einem speziellen Proportional-Szintillationsdetektor beruht. Bei mäßiger erreichbarer Meßgenauigkeit kommen dabei Vorteile wie Schnelligkeit und der nichtzerstörende Charakter solcher röntgenradiometrischer Verfahren voll zum Tragen. Verwendet wird ein RFA-Meßsystem, das unter der Bezeichnung ,,Röntgenradiometrischer Mehrkanalanalysator für leichte Elemente mit Proportional-Szintillations-Detektor" bereits eingehend beschrieben wurde [2, 3]. Insbesondere eine sowjetische Version dieses Meßsystems ist das Ergebnis einer langjährigen, zu beiderseitigem Nutzen geführten und sich weiter planmäßig fortsetzenden Zusammenarbeit unseres Institutes mit der Moskauer Partnereinrichtung Allunions-Forschungsinstitut für Strahlentechnik des Staatlichen Komitees für Kernenergie der UdSSR.

2. Beschreibung des Meßverfahrens und experimentelle Ergebnisse

Das hier zu beschreibende Verfahren der Ermittlung des O/U-Verhältnisses stellt also eine quantitative Bestimmung der beiden Komponenten U und O mit den Mitteln der energiedispersiven Röntgenfluoreszenzanalyse dar. Dabei bedingt die als ,,ultraweich" zu bezeichnende Energieregion der bei 525 eV gelegenen K-Strahlung des Sauerstoffs, die zur Auswertung dieser Komponente allein zur Verfügung steht, und der unmittelbar benachbarten N-

[1]) Anschrift: DDR-7050 Leipzig, Permoserstr. 15

Strahlung des Urans (ca. 300 eV) eine Anregung der Röntgenstrahlung mit schweren geladenen Partikeln, z.B. α-Strahlung des Nuklides ^{238}Pu, und das Arbeiten in einer evakuierbaren Meßkammer.

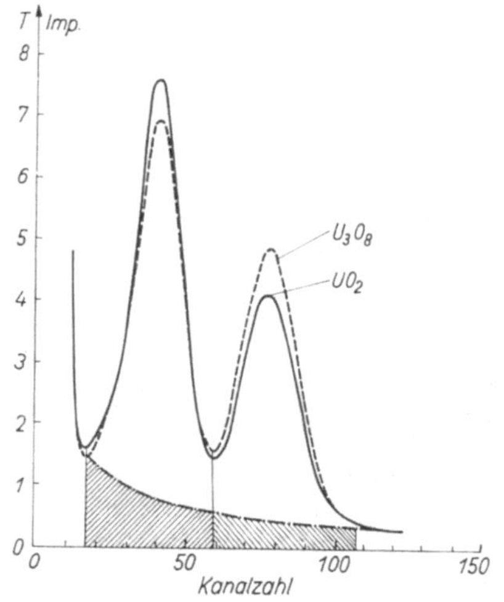

Abb. 1. Spektrogramm der U-N- und O-K-Strahlung für UO_2 und U_3O_8

Wie das Spektrogramm in Abb. 1 anschaulich demonstriert, ist die zum Nachweis dieser ultraweichen Röntgenstrahlung verwendete sogenannte Kugelvariante des Proportional-Szintillations-Detektors infolge seines in diesem Gebiet von bisher keiner anderen Detektorart übertroffenen Energieauflösungsvermögens in der Lage, die beiden Strahlungen nahezu vollständig zu trennen. Auf eine eingehende Beschreibung des Aufbaues und der Wirkungsweise des Proportional-Szintillations-Detektors und des gesamten Analysators kann hier verzichtet werden (siehe [2]). Es sei nur darauf hingewiesen, daß im Sinne des Erzielens eines möglichst günstigen Berg-Tal-Verhältnisses, insbesondere auch einer weitestgehenden Abtrennung des U-N-Peakes von dem intensiven niederenergetischen Untergrund, für die vorliegenden Untersuchungen eine spezielle Zählgasmischung aus 80% Xe und 20% He verwendet wurde, die sich insofern als optimal erwies, als eine möglichst vollständige Sammlung der primären Ladungsträgerpaare erreicht und somit eine starke Ausprägung von ,,tailings" auf der niederenergetischen Peakseite eingeschränkt werden konnte. Andererseits behält die mit zunehmendem He-Gehalt an sich sinkende Lumineszenzausbeute des Detektionsgases noch vertretbare Werte.

Für eine Kompensation eventuell auftretender Schwankungen des Detektor- bzw. Verstärker-Ausgangssignales sorgt ein digitaler Impulshöhenstabilisator [4]. Dieser benötigt eine nach Möglichkeit im Meßspektrum enthaltene, isoliert stehende und bei schwankenden Gehalten ihre Form nicht ändernde Referenzlinie, wofür sich hier die bei etwa 3,3 keV gelegene M-Strahlung des Urans anbietet (vgl. Abb. 2).

Unter dem Blickwinkel zu erwartender Störungen und Verfälschungen stellt die Substanz UO_{2+x} ein geradezu ideales Untersuchungsobjekt dar, denn im ultraweichen Röntgengebiet an sich verstärkt wirksame Interelementeffekte sind bei einem solchen Zweistoffsystem nicht zu befürchten.

Das zu untersuchende Probenmaterial lag als UO_2-Pulver und auch in Form gesinterter zylinderförmiger Pellets (Länge ca. 11 mm, Durchmesser 7,5 mm) vor. Das Pulvermaterial, das eine spezifische Oberfläche von

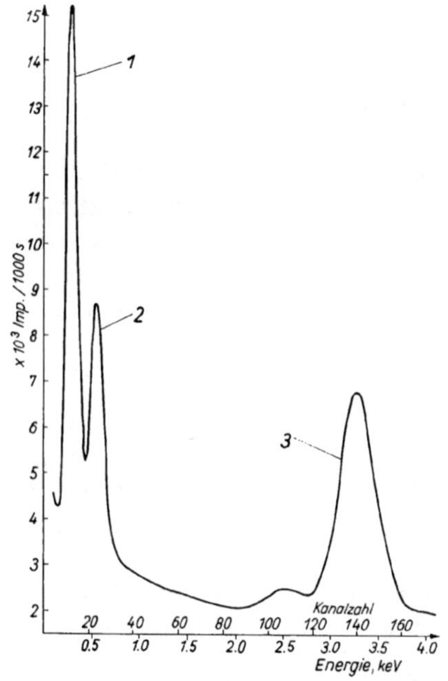

Abb. 2. Spektrogramm der U-N-, O-K- und U-M-Strahlung
1 – U-N-Strahlung; 2 – O-K-Strahlung; 3 – U-M-Strahlung

4,4 m²/g aufwies, wurde bei etwa 30 kp/cm² zu Tabletten mit einem Durchmesser von 20 mm verpreßt, deren Haltbarkeit auch unter den Vorvakuum-Bedingungen ohne Zugabe von Bindemittel durch die Verwendung von Aluminium-Trägerringen gewährleistet war. Ein gegenüber [2] veränderter Proben-Wechsel- und -Halter-Mechanismus gestattete das definierte Einbringen sowohl dieser Tabletten als auch von Pellets (einzeln oder in ,,Paketen") in die evakuierbare Bestrahlungs- und Meßkammer.

Als primäre α-Strahlungsquelle kam ein auf der Basis von gesintertem keramischen Trägermaterial hergestelltes Präparat des Nuklides ^{238}Pu zum Einsatz. Diese vom sowjetischen Partner zur Verfügung gestellte Quelle vom Typ AIPRA war ringförmig aufgebaut und besaß eine Aktivität von ca. 0,4 GBq. Sie hat sich, nachdem zum Zwecke des Zurückhaltens von Photo- und Konversions-Elektronen eine zusätzliche Goldschicht von ca. 3 μm Dicke durch Aufdampfen aufgebracht worden war, bezüglich ihrer Emissionseigenschaften und über Jahre hinweg auch hinsichtlich ihrer mechanischen Stabilität und Kontaminationssicherheit hervorragend bewährt.

Die praktischen Messungen wurden nun in der Weise vorgenommen, daß mittels zweier elektronischer Kanäle – realisiert durch zwei ,,region of interest" bei Nutzung eines Vielkanalanalysators, sonst durch Verwendung zweier Einkanalanalysatoren – die Intensitäten der O-K- bzw. U-N-Peaks parallel ermittelt wurden (Abb. 1). Es hat sich als zweckmäßig erwiesen, als Maß für das O/U-Verhältnis den Quotienten Q aus diesen beiden Intensitätswerten zu bilden, um damit von elektronischen Stabilisierungsmaßnahmen nicht zu erfassende Gänge der Apparatur oder auch matrixeffektähnliche Störungen dank der energetischen Nachbarschaft der Strahlungen weitestgehend zu kompensieren.

Zur Aufstellung einer Eichkurve standen zunächst nur zwei Substanzen zur Verfügung, deren O/U-Verhältnisse genügend genau bekannt und nahe bei denen der Verbindungen UO_2 und U_3O_8 gelegen waren. Da sich jedoch der Masseanteil beim Übergang von UO_2 zu U_3O_8 um jeweils nur ca. 3,4% ändert, kann zusammen mit der Tatsache des idealen Zweistoffsystems ein linearer Verlauf der Eichkurve für den Quotienten Q angenommen werden.

Zur Präparierung der Eichproben stand ein pulverförmiges Urandioxid mit einer spezifischen Oberfläche von 4,4 m^2/g zur Verfügung, das an Luft über einen Zeitraum von 1,2 Jahren bei einer mittleren Temperatur von 20 °C gelagert worden war. Gemäß *Bannister* [5] führen diese Ausgangsdaten infolge Oxidation zu einem sich selbständig einstellenden O/U-Verhältniss von 2,14.

Der zweite Meßpunkt wurde durch eine definierte Oxidation dieses Pulvers zu U_3O_8 erhalten, wobei auf „klassischem" Wege ein O/U-Verhältnis von 2,694 ermittelt wurde. Die somit erhaltene Eichkurve zeigt Abb. 3, auf der die erhaltenen, mit einem apparaturbedingten Faktor versehenen Quotienten Q aus den Impulszahlen im O- und U-Kanal gegen die genannten O/U-Werte aufgetragen wurden.

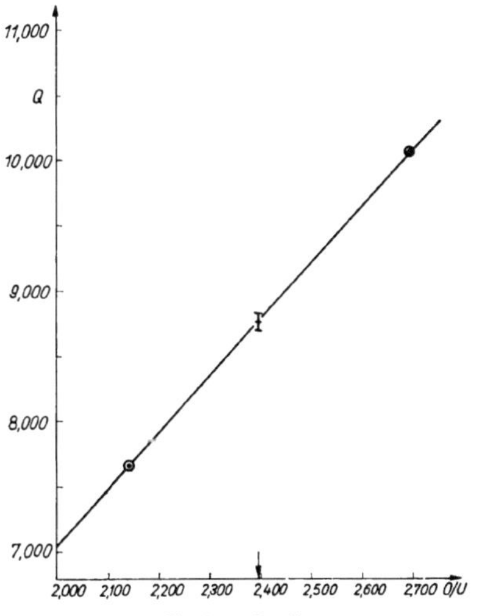

Abb. 3. Eichkurve für die O/U-Bestimmung

Ein Aliquot der Ausgangssubstanz wurde nunmehr einer über insgesamt 5 Tage dauernden thermischen Behandlung an Luft bei einer Temperatur von 98 °C unterzogen. Das Ausmessen einer aus diesem Uranoxidpulver hergestellten Tablette erfolgte dann über einen Zeitraum von über 8 h und erbrachte eine Meßreihe von 50 Intensitätsverhältniswerten, deren Mittel zu 8,742 mit einem mittleren quadratischen Fehler der Einzelmessung von $\sigma = \pm 0{,}036$ bestimmt wurde. Dieser Mittelwert ergibt gemäß Eichkurve (Abb. 3) ein O/U-Verhältnis von 2,395. Er besitzt den Charakter sowohl eines Meßpunktes für eine Substanz mit „unbekanntem" O/U-Verhältnis als auch eines weiteren Stützpunktes für die Eichkurve, wie sich aus den folgenden Überlegungen ergibt:

Scheibe u. a. [6] berichten über das Reoxidationsverhalten von Urandioxid, wenn dieses über längere Zeit (mehrere Tage) Temperaturen im Bereich 80 bis 110 °C ausgesetzt war. So stellte sich bei einer thermischen Behandlung an Luft bei 80 °C nach ca. 3,5 Tagen ein O/U-Sättigungswert von 2,33 ein, was mit ziemlicher Sicherheit einer Oxidationsstufe U_3O_7 zugeschrieben werden kann. Eine Erhöhung der Temperatur auf 110 °C führte hingegen zu einem Sättigungswert von 2,40, was einem nahezu vollständigen Übergang des Uranoxides in die Form U_5O_{12} entspricht. Das Ergebnis von O/U = 2,395, das bei einem mehr zu höheren der beiden Behandlungstemperaturen tendierenden Temperaturwert von nahezu 100 °C von uns erhalten wurde, steht mit diesen Vorstellungen in Einklang und paßt sich auch in die durch bisher zwei Stützpunkte aufgespannte Eichgerade geradezu ideal ein.

Bei der Ermittlung des zu erwartenden Meßfehlers einer O/U-Wert-Messung wird von einer für eine solche Expreß-Bestimmung typischen Meßzeit der Größenordnung 10 min ausgegangen. Damit ergeben sich im Mittel 10^5 Impulse für die Intensität im U-Kanal (I_u) und $8{,}6 \cdot 10^4$ Impulse für die im O-Kanal (I_0), für die zählstatistischen Standardabweichungen entsprechend $\Delta I_u = 316$ und $\Delta I_0 = 293$.

Ermittelt man nun gemäß

$$\Delta Q = \frac{\Delta I_0}{I_u} + I_0 \frac{\Delta I_u}{I_u^2} \tag{1}$$

den bei der Quotientenbildung auftretenden Fehler ΔQ und rechnet weiter die Steigung der Eichgeraden ein, so ergibt sich ein O/U-Meßfehler von

$$\Delta(O/U) = \pm 0{,}01 \text{ O/U-Einheiten}.$$

Erwähnenswert erscheint, daß ein $\Delta Q = \pm 0{,}03$ gemäß Gleichung (1) bemerkenswert gut mit dem Wert $\sigma = \pm 0{,}036$ korrespondiert, der sich bei der Auswertung der o. g. Meßreihe ergeben hatte.

3. Schlußfolgerungen

Zwar liegen die erreichbaren Meßfehler einer solchen röntgenradiometrischen Methode für die Bestimmung des O/U-Verhältnisses um mindestens eine Größenordnung über denen anderer, „klassischer" Bestimmungsverfahren, jedoch sollten die Vorteile einer kurzen Meßdauer, der Zerstörungsfreiheit und damit Wiederverwendbarkeit des UO_2-Materials ohne jede Veränderung des chemischen, im Falle von Pellets auch des physikalischen Zustandes dieses Meßverfahren geeignet erscheinen lassen, in kurzer Zeit eine grobe Orientierung zu erlangen und einen hohen routinemäßigen Probendurchsatz zu ermöglichen.

Die bei der Anregung mit schweren geladenen Teilchen auftretende Besonderheit der Erfassung einer relativ dünnen Oberflächenschicht stellt im Falle der Herstellung von Tabletten keine Einschränkung dar, wenn das Probenmaterial für die Tabletten repräsentativ für eine bestimmte Stoffmenge ist. Im Falle der Pellets kann andererseits diese Besonderheit sogar von Vorteil sein, wenn eine eventuelle Sauerstoffanreicherung einen gerade in oberflächennahen Schichten auftretenden und deshalb mit anderen Verfahren nur schwer erfaßbaren Effekt darstellt, weil diese i. allg. einen mittleren Wert über eine mehr oder weniger große Stoffmenge liefern.

So ist eine Verfolgung der Kinetik eines oberflächennahen Oxidationsvorganges am gleichen Objekt über einen langen Zeitraum denkbar, ober aber es lassen sich, wenn man an ein Abtragen definierter Oberflächenschichten mit mechanischen oder chemischen Mitteln und einer anschließenden O/U-Bestimmung an der dann freiliegenden Oberflächenschicht denkt, Tiefenprofile der O/U-Werte etwa von UO_2-Pellets herstellen.

Eingegangen am 15. 4. 1985

Literatur

[1] *H. Nickel*, Nukleonik 8 (1966) 366
[2] *D. Müller, P. Morgenstern, W. Riedel, N. A. Anders, A. A. Waigacev, W. P. Warwariza, W. S. Isajev, B. E. Kolesnikov, S. W. Mamikonjan, W. I. Filatov, J. M. Schokin*, Isotopenpraxis **17** (1981) 12
[3] *N. A. Anders, W. P. Warwariza, B. E. Kolesnikov, S. W. Mamikonjan, L. P. Starschik, W. I. Filatov*, Fragen der Atomwissenschaft und -technik, Serie Strahlentechnik, **1** (1982) No. 23, p. 125
[4] *E. Dahn, W. Riedel, M. Braune*, Isotopenpraxis **3** (1977) 135
[5] *M. J. Bannister*, AAEC/TM 396 (July 1967)
[6] *H. Scheibe, W. Ermischer*, Kernenergie **7** (1964) 95

Isotopenpraxis 22 (1986) 2, S. 68 bis 69

Schnelle Bestimmung von Arsen in Erzen und Aufbereitungsprodukten mit der Radionuklid-Röntgenfluoreszenzanalyse*)

G. Krüger

(VEB Mansfeld-Kombinat „Wilhelm Pieck", Forschungsinstitut für NE-Metalle Freiberg)[1])

Es wird eine Verfahrensvariante vorgestellt, die unter Nutzung eines weiterentwickelten rechnergestützten Gerätes vom Typ ITRA (Röntgenfluoreszenzanalysator mit Nuklidanregung) die schnelle Bestimmung von Arsen in Erzen und Aufbereitungsprodukten ermöglicht. Dabei wird auf die Beeinflussung der Meßergebnisse durch unterschiedliche Matrixzusammensetzung eingegangen. Eine Möglichkeit, wichtige Begleitelemente mit zu analysieren und zur Korrektur zu verwenden, wird aufgezeigt.

In this paper a technique is described which by the use of an improved computerized device type ITRA (X-ray fluorescence analyzer with nuclide excitation) allows rapid determination of arsenic in ores and dressing products to be performed. The influence of different matrix composition on the results is discussed. The possibility of determining important matrix elements that can be used to make corrections is suggested.

Keywords

computers; corrections; errors; filters; scintillation counters; X-ray fluorescence analysis

1. Einleitung

Die Notwendigkeit der komplexen Nutzung einheimischer Rohstoffe und Probleme der Umweltbelastung erfordern schon bei der Erzaufbereitung gezielte Maßnahmen. Die optimale Separierung getrennt zu gewinnender oder im nachfolgenden Verhüttungsprozeß störender Minerale bzw. Elemente spielt dabei eine große Rolle. Diese Aufgaben stellen an die Betriebslaboratorien erhöhte Anforderungen, die mit dem bestehenden Arbeitskräftefonds nicht oder nur schwer zu bewältigen sind.

Hochmoderne Analysenautomaten können in vielen Fällen diese Probleme lösen, doch erfordert ihr Einsatz erhöhte personelle und materielle Aufwendungen, die nur vereinzelt gerechtfertigt und zu realisieren sind.

Das Forschungsinstitut für NE-Metalle bemüht sich seit einigen Jahren mit Erfolg durch Entwicklung und Einsatz entsprechender Radionuklid-Röntgenfluoreszenzanalysatoren um die Schließung dieser Lücke [1].

2. Aufgabenstellung

Es besteht die Notwendigkeit, Arsengehalte in Erzen und Aufbereitungsprodukten schnell zu bestimmen. Dabei ist zu berücksichtigen, daß größere Eisengehalte sowie Gehalte von Wolfram und Bismut die Messung beeinflussen können. Unter Einbeziehung der gerätetechnischen Möglichkeiten, die ein verfügbarer Röntgenfluoreszenzanalysator bietet, war ein Verfahren auszuarbeiten, das den gestellten Forderungen gerecht wird.

3. Durchführung der Arbeit

3.1. *Gerätetechnik*

In Fortführung bisheriger Arbeiten [1, 2] wurde auf der Basis handelsüblicher Geräte (Einkanalspektrometer, Szintillationszähler) ein Meßplatz entwickelt und aufgebaut, der, durch entsprechende Vorwahlmöglichkeiten und Mikrorechnersteuerung, Messungen nacheinander in vier verschiedenen Energiebereichen (Kanälen) bei vier verschiedenen Filterpositionen ermöglicht, wobei die Auswertung der Meßergebnisse automatisch erfolgt [3]. Somit sind verschiedene Meßverfahren möglich, doch gelangt, um das relativ schlechte Auflösungsvermögen des Szintillationszählers auszugleichen, im wesentlichen das Kantenfilter-Differenzverfahren zur Anwendung.

Die Anregung der Röntgenfluoreszenzstrahlung erfolgt mit einer ^{238}Pu-Strahlenquelle (1,11 GBq) mit Berylliumfenster in Zentralgeometrie. Der Szintillationszähler wurde mit einem 2 mm dicken NaI(Tl)-Kristall ausgerüstet.

3.2. *Meßverfahren*

In Abb. 1 sind die Linien von Elementen, die für eine Arsenbestimmung von Interesse sind, und Kanten von Materialien, die als Filter in Frage kommen können, zusammengestellt.

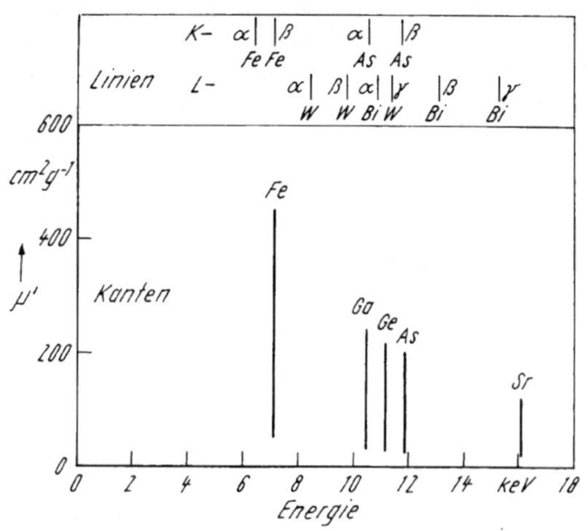

Abb. 1. Zusammenstellung von Linien und Kanten der interessierenden Elemente zur Filterauswahl

*) Vortrag, gehalten zum 7. Kolloquium „Technische Anwendung radioaktiver Isotope im Berg- und Hüttenwesen", 13. bis 14. März 1985, Hetzdorf, DDR

[1]) Anschrift: DDR-9200 Freiberg, Lessingstraße 41

Der relativ große Energieunterschied zwischen den emittierten Linien, der in der Probe angeregten Elemente, macht die Auswahl der benötigten Filtermaterialien dann unproblematisch, wenn solche Begleitelemente wie W, Bi und Pb fehlen oder in konstantem Anteil in der Probe vorhanden sind.

Im vorliegenden Fall mußte neben dem zu bestimmenden Arsen mit schwankenden Gehalten an Eisen, Wolfram und Bismut gerechnet werden. Je nach Produkt (Erz, Zwischenprodukte, Konzentrat) unterscheiden sich die Gehalte der Elemente in ihrem absoluten Wert aber auch in ihrer Schwankungsbreite.

Entsprechend der gegebenen gerätetechnischen Möglichkeiten wurden 4 Filter ausgewählt (Fe, Ga, As, Sr), mit deren Hilfe es durch entsprechende Kombination möglich ist, alle in dem zu untersuchenden Material interessierenden Elemente zu erfassen. Da jeweils zwei Filter für die Erfassung der Strahlung innerhalb eines bestimmten Energiebereichs verwendet werden, somit ein Filter für die Messung der angeregten Strahlung in zwei Energiebereichen zum Einsatz gelangt, bereitet die Abstimmung besondere Schwierigkeiten.

Die tatsächlich verwendeten Flächenmassen der Filter können daher von denen, die als optimal errechnet wurden (siehe Tab. 1), erheblich abweichen.

Tab. 1. Optimale Flächemasse von Filtern (nach [4])

zu bestimmendes Element	Linie	Energie keV	Filterelement	Flächenmasse des Filters mg/cm²
Bi	L_β	13,02	Sr	10,1
			As	14,2
As	K_α	10,54	As	7,7
			Ga	10,6
	K_β	11,72	As	10,3
			Ga	14,3
W	L_α	8,40	Ga	4,9
			Fe	8,5
	L_β	9,67	Ga	7,4
			Fe	12,8

Für die einzelnen Elemente wurden unabhängig von ihrer wechselseitigen Beeinflussung folgende Meßgrößen ausgewählt:

$I_W \triangleq F2 - F1$ $\quad F1$ = Impulsrate bei Filter 1 (Fe)
$I_{As} \triangleq F3 - F2$ $\quad F2$ = Impulsrate bei Filter 2 (Ga)
$I_{Bi} \triangleq F4 - F3$ $\quad F3$ = Impulsrate bei Filter 3 (As)
$F4$ = Impulsrate bei Filter 4 (Sr)

Da nur 4 Filterplätze zur Verfügung stehen, wurde als Information über den Fe-Gehalt der Probe das Verhältnis Nutzpeak/Streustrahlung bei Filter 1 (durchlässig für die Fe-Strahlung) durch Umschaltung des Meßkanals ermittelt.

$$I_{Fe} \triangleq \frac{F1\,K1}{F1\,K2}$$

K 1 = Messung im Kanal 1 (Nutzpeak)
K 2 = Messung im Kanal 2 (Streustrahlung)

Die gegenseitige Beeinflussung der einzelnen Meßgrößen macht eine Auswertung an einem externen Rechner erforderlich. Dabei wird der sog. Potenzreihenansatz zugrunde gelegt [5]. Aus dem vorgegebenen quadratischen Ansatz mit allen Mischgliedern ergeben sich unter Anwendung des Signifikanztestes für die einzelnen Elemente reduzierte Auswertefunktionen, die z.B. für die vorliegende As-Bestimmung folgende Form besitzt:

$$\% \text{ As} = a_0 + a_1 I_{As} + a_2 I_{Bi} + a_3 (I_{Fe})^2 + a_4 (I_W)^2 + a (I_{As})^2 \quad (1)$$

Nach Eingabe des entsprechenden Meßprogramms und der Koeffizienten a_i in den Rechner des RF-Analysators läuft der Meßvorgang bis zur Gehaltsausgabe automatisch ab.

4. Ergebnisse

Verfahren der Röntgenfluoreszenzanalyse sind eichbedürftig. Somit wird die Güte des Verfahrens sehr wesentlich durch die Verfügbarkeit entsprechend richtig analysierter Eichproben bestimmt. Besonders brisant wird dieses Problem dann, wenn, wie bei der zu lösenden Aufgabe, neben dem eigentlich zu bestimmenden Element andere Begleitelemente die Messung wesentlich beeinflussen und es erforderlich ist, daß auch der Gehalt dieser Begleitelemente sehr präzise ermittelt wird.

Im vorliegenden Fall konnte auf eine Testreihe von Erzproben zurückgegriffen werden (siehe [2]), die im Zentralinstitut für Kernforschung Rossendorf mittels Neutronenaktivierungsanalyse untersucht worden war. Aus dieser Testreihe wählten wir 10 Proben aus, die die uns interessierenden Elemente in entsprechenden Konzentrationen enthielten. Für den Bi-Gehalt lagen keine Analysenwerte vor, deshalb wurden in die Berechnungen die ermittelten Impulsraten eingesetzt.

In Tab. 2 sind für die Elemente Fe, W und As die Konzentrationsbereiche, die durch die 10 Proben repräsentiert wurden, und die ermittelte Reststreuung angegeben. Die Ergebnisse sind durchaus mit den in den Betriebslabors üblichen Werten vergleichbar. Nach Vorliegen der Analysenwerte für Bismut kann die vorgeschlagene Methode um die Bi-Bestimmung erweitert werden.

Tab. 2. Meßergebnisse

zu bestimmendes Element	Gehaltsbereich %	Reststreuung s_R abs. %
Fe	3 ... 30	1,22
W	0 ... 0,6	0,02
As	0 ... 1,2	0,05

5. Schlußfolgerung

Röntgenfluoreszenzanalysatoren mit Isotopenanregung sind auch in der Grundstoffindustrie der DDR eine wichtige Ergänzung konventioneller Analysentechnik. Sie können dort effektiv eingesetzt werden, wo relativ schnell Analysenergebnisse einiger weniger Elemente für die Produktionssteuerung benötigt werden. Durch wesentliche Verkürzung der Analysendauer stellen sie in bestimmten Fällen auch ein effektives Rationalisierungsmittel dar. Durch den Einsatz der Mikrorechentechnik werden sie sich weitere Anwendungsgebiete erschließen.

Eingegangen am 14. 5. 1985

Literatur

[1] *E. Lehmann, G. Krüger*, Isotopenpraxis **15** (1979) 179

[2] *G. Krüger, W. Böhme, M. Thomae, G. Loos*, Isotopenpraxis **19** (1983) 198

[3] *V. Kliem*, Vortrag zum 7. Kolloquium „Technische Anwendung radioaktiver Isotope (TARI)“ Hetzdorf 1985

[4] *C. Broquet*, Chimie Analytique **52** (1970) 996

[5] Autorenkollektiv, Röntgenfluoreszenzanalyse – Anwendung in Betriebslaboratorien. VEB Deutscher Verlag für Grundstoffindustrie, Leipzig 1981, S. 106 ff.

Isotopenpraxis 22 (1986) 2, S. 70 bis 72

Buchbesprechungen

A Concept of Mathematical Physics
Models for Space-Time

T. Matolcsi. Budapest: Akadémiai Kiadó 1984. 235 S., div. Abb.; kart. M 67,50

Dieses Buch ist als ein Ansatz zu verstehen, die Grundbegriffe der Physik auf eine solide mathematische Basis zu stellen. Nach Meinung des Autors — der man sich im wesentlichen anschließen kann — befindet sich die theoretische Physik, speziell die Quantenfeldtheorie, in einer Krise, die der der Mathematik gegen Ende des vorigen Jahrhunderts vergleichbar ist und die aus der mangelnden mathematischen Exaktheit in der Definition der physikalischen Grundbegriffe resultiert. Gefordert wird deshalb eine neue Theorie — die mathematische Physik —, die insbesondere den Begriff des Modells auf streng mathematischer Grundlage axiomatisch fundiert.

Das Buch widmet sich in seinem ersten Teil der Konstruktion mathematischer Modelle für das Raum-Zeit-Kontinuum, wobei nichtrelativistische, speziell- und allgemeinrelativistische Raum-Zeit-Modelle axiomatisch konstruiert werden. Abweichend von den allgemeinen Vorstellungen wird dabei schon im nichtrelativistischen Fall eine orientierte, affine Struktur der Raum—Zeit angenommen, wobei aber noch eine absolute Zeit existiert. Diese Einschränkung wird dann in den relativistischen Modellen aufgehoben.

Der zweite Teil des Buches stellt die mathematischen Grundlagen dar. Behandelt werden Tensoroperationen, pseudo-euklidische und affine Räume, glatte Mannigfaltigkeiten und *Lie*sche Gruppen, wobei der axiomatische Aspekt besonders betont wird.

Das Buch setzt eine gute mathematische Grundbildung voraus und ist besonders Spezialisten der Relativitätstheorie und Quantenfeldtheorie zu empfehlen. *R. Der*

Lehrbuch der Statistik
Für Sozialwissenschaftler

J. Bortz. 2. vollst. neubearb. u. erw. Aufl.; Berlin-Heidelberg-New York-Tokyo: Springer-Verlag 1985. 898 S., 71 Abb., 223 Tab.; geb. DM 78.—

Das Lehrbuch ist in drei große Teile gegliedert: Elementarstatistik, varianzanalytische und multivariate Methoden, die jeweils ungefähr 250 Seiten ausmachen, und jeder dieser Teile ist in 6 überschaubare Kapitel untergliedert. So finden sich im 1. Teil erwartetermaßen die verschiedenen Verteilungen, die Stichprobenstatistik, statistische Prüfverfahren und die lineare Regression. Im 2. Teil werden die ein- und mehrfachen Versuchspläne und die Kovarianzanalyse behandelt und im letzten Teil die multiple Regression, die Faktorenanalyse, die Cluster- und Diskriminanzanalyse. Jedem dieser Kapitel sind Übungsaufgaben zugeordnet, die am Schluß des Lehrbuches in einem Anhang aufgelöst werden. Ein Tabellenanhang, ein umfangreiches Literaturverzeichnis und ein Namens- und Sachverzeichnis schließen das Buch ab.

Dieser offensichtlich bewährte Aufbau des Lehrbuches hat noch weitere Vorzüge: die ausführliche Darstellung des Rechenganges an den vielen ausgewählten Beispielen und der sorgfältige Umgang mit den erhaltenen Zahlenergebnissen der statistischen Methoden, die nicht leichtfertig zu weitreichenden Schlußfolgerungen verlocken sollten. So werden das Verfahren der Nullhypothese (oder Alternativhypothese), die einfache und mehrfache Korrelation oder das hypothesengenerierende Verfahren der Faktorenanalyse mit den nötigen Hinweisen zu ihrer Interpretation versehen — ein Vorzug, aber auch eine zwingende Notwendigkeit, um als Lehrbuch wirken zu können. Gerade in einer Zeit, in der die teils aufwendige Rechentechnik der statistischen Verfahren leicht durch Taschenrechner oder Bürocomputer zu Zahlenwerten führen, ist ihre Interpretation und ihr Aussagefeld indispensibel, denn die statistischen Verfahren sollen Beweiskraft und nicht Alibicharakter für unbewußt erwünschte Zusammenhänge haben. Und daß gerade das dargestellte Anwendungsgebiet der Statistik, die Sozialwissenschaften, ganz besonders häufig auf nicht einfache lineare Weise die Wirklichkeit widerspiegeln, ist leicht vorstellbar.

Die Literatur ist für ein Lehrbuch sehr umfangreich zitiert — insgesamt 734 Literaturstellen —, wobei seit der 1. Auflage aus dem Jahre 1977 bis 1984 allein 192 ($\triangleq$ 26,2%) Literaturstellen hinzugekommen sind. Damit ist dem Lernenden ein breites Angebot zur Vertiefung und Erweiterung des dargebotenen Wissens gegeben, was gewiß öfters in Anspruch genommen werden wird. *H. Hübner*

Messung ionisierender Strahlung

W. Stolz. Berlin: Akademie-Verlag 1985. 190 S., 53 Abb., 17 Tab. (WTB Mathematik-Physik, Bd. 297); M 12,50

Der Autor konstatiert: „Quellen ionisierender Strahlung finden in vielen Bereichen der Naturwissenschaft, Technik, Medizin und Landwirtschaft Anwendung. Die Grundvoraussetzung dafür bildet die Strahlungsmeßtechnik". Damit ist die Aufgabenstellung des Buches deutlich abgegrenzt. Da die elektronischen Verfahren der Meßwerterfassung und -verarbeitung ebenfalls ausgeklammert sind und die Kapitel Strahlungsquellen und Strahlungsgrößen sowie Dosismessung nur knapp dargestellt werden, gelingt auf 150 Seiten eine in sich geschlossene und gut verständliche Darstellung der physikalischen Grundlagen des Nachweises ionisierender Strahlung, der Funktion und Ausführungsform der für die Praxis wichtigsten Strahlungsdetektoren sowie der häufigsten, grundlegenden Meßprobleme.

In leicht verständlicher Form werden Ionisationskammern, Proportional- und Auslösezählrohre, Halbleiterdetektoren, Szintillationszähler, *Čerenkov*-Zähler, Photographische Emulsionen, Aktivierungssonden, Festkörperspurdetektoren und chemische Detektoren beschrieben. Bei den Meßproblemen findet man z.B. Verfahren zur Aktivitätsmessung, die α-, β-, γ-, n-Spektrometrie und die Messung von Halbwertszeiten.

Der Autor hat das Buch insbesondere für Studierende naturwissenschaftlich-technischer Fachrichtungen niederer Semester geschrieben. Es ist darüber hinaus geeignet, all denen, die in Naturwissenschaft, Technik, Medizin und Landwirtschaft den Einsatz von Methoden und Verfahren auf Basis ionisierender Strahlung erwägen, einen Überblick über verfügbare Methoden und Geräte zur Strahlungsmessung, ihrer Arbeitsweise und Leistungsparameter zu geben.

Wegen der strengen thematischen Eingrenzung des Buches ist in vielen Fällen das Studium allgemeiner wie weiterführender Literatur unumgänglich. Der Autor gibt im Literaturverzeichnis dafür sehr zweckmäßige Hinweise, freilich auch auf gute umfassendere Darstellungen, die den Inhalt des vorgelegten Buches, eingeordnet in die Gesamtproblematik, in etwa mit enthalten. *G. Hartmann*

Nutzung der Uraniumisotope ^{234}U und ^{238}U in der Hydrologie und Hydrogeologie

R. Gellermann, K. Fröhlich. Leipzig: VEB Deutscher Verlag für Grundstoffindustrie 1984. 101 S., 35 Abb., 20 Tab. (Freiberger Forschungshefte C 397 Geowissenschaften/Geologie); geheftet M 31,—

Es sind nur wenige Fälle bekannt, in welchen ein Isotopenverhältnis durch die Messung der Aktivitäten bestimmt werden kann. Es setzt die Radioaktivität beider beteiligter Isotope voraus und soll natürlich getrennt voneinander meßtechnisch erfaßbar sein. Die Anwendbarkeit solcher Isotopenverhältnisse in den Geowissenschaften ist zusätzlich an die Bedingung geknüpft, daß die Messungen im natürlichen Bereich möglich sind und daß je nach Vorgeschichte auch deutliche Unterschiede in den untersuchten Proben nachgewiesen werden können. Man erkennt leicht, daß eine Reihe von Voraussetzungen erfüllt sein müssen, bevor ein weiteres Isotopenpaar als moderner Indikator für wissenschaftliche Fragen einsetzbar ist. Das Uranium-Isotopenpaar ^{234}U und ^{238}U erfüllt in vollem Umfang diese genannten notwendigen und hinreichenden Bedingungen.

Das kleine Bändchen der bekannten Freiberger Forschungshefte ist neben einer Einleitung, einem Formelzeichenverzeichnis und einem Literaturverzeichnis in 4 große Abschnitte gegliedert: Messung der Uraniumisotope, Isotopenfraktionierungen der U-Isotope in natürlichen Wässern, Methoden der Nutzung von U-Isotopen in der Hydrologie und Hydrogeologie und einige ausführlich beschriebene Anwendungsbeispiele. Dabei stellen die Autoren einerseits den Stand des Wissens auf diesem Gebiet anhand einer sorgfältigen Literaturauswertung dar, und andererseits werden einige eigene Arbeiten und die erhaltenen Ergebnisse vorgestellt und diskutiert.

Obwohl die Erscheinung der Isotopenvariationen des $^{234}U/^{238}U$-Verhältnisses 1955 zum ersten Mal von *V. V. Cherdyntsev* beschrieben wurde, ist seither eine bemerkenswerte Literatur darüber erschienen, die mit 139 Zitaten belegt wird, wobei allein aus den Jahren 1976—1984 rund 62% der Zitate stammen; ein Hinweis, daß bis in die jüngste Zeit diesem Phänomen Aufmerksamkeit geschenkt wird. Obwohl die Tiefseesedimente nicht Gegenstand der Darstellung waren, hätte man auf die hohen Werte des Aktivitätsverhältnisses der Mn-Knollen aufmerksam machen können, die vielleicht in einer ozeanischen Bilanzierung mit berücksichtigt werden müßten.

Wie bei allen Isotopenverfahren, so ist auch die Anwendung dieses Isotopenverhältnisses an zusätzliche unabhängige Parameter gebunden, was von den Autoren an mehreren Stellen deutlich gemacht wird, Es ist eine Schrift, die eine Zusammenstellung der möglichen Anwendungen in der Hydrosphäre anbietet, wofür nicht nur der Isotopenhydrologe sein hohes Interesse anmelden wird, sondern auch der Praktiker der Wasserwirtschaft, um sich zu informieren und vielleicht auch diese Methode an geeigneter Stelle anzuwenden. *H. Hübner*

Mathematische Modellierung von Prozessen in medizinischen und biologischen Systemen

Sammelband. Hrsg.: V. A. Bajdosov, L. E. Fefelova. Sverdlovsk: Akademija Nauk SSSR (Ural'skij Naučnyj Centr) 1982. 80 S., 18 Abb., 7 Tab., 61 Lit. (in Russisch); brosch. Rubel 0,85; M 4,25

Das Buch vereinigt 10 Arbeiten von Wissenschaftlern des Instituts für Mathematik und Mechanik am Wissenschaftlichen Zentrum des Urals der Akademie der Wissenschaften der UdSSR in Sverdlovsk und spiegelt die Vielfalt der Aktivitäten dieses Instituts auf dem Gebiet der mathematischen Modellierung biologischer Prozesse wider. Behandelt werden zur Einführung methodische Probleme, weiterhin die Kinetik von radioaktivem Iod (^{131}I) im menschlichen Organismus (lineare Compartmentmodelle), die Regulation der Ausscheidung toxischer Metalle aus dem Organismus (Modellierung durch nichtlineare Differentialgleichungen), ferner Fragen der Radionukliddiagnostik und der Prognose pathologischer Zustände bei der Geburt unter Nutzung eines am Institut entwickelten Rechenprogramms zur Mustererkennung, die Automatisierung von Diagnoseverfahren in der Kardiologie und schließlich die mathematische Modellierung der genetischen Struktur von Populationen und des Entwicklungsprozesses von Zellen unter Extremalbedingungen, insbesondere bei der Einwirkung ionisierender Strahlung (Wiedergabe des Prozeßablaufs durch lineare Differentialgleichungen).

Die Art und Weise der mathematischen Behandlung der betrachteten Vorgänge ist ausgewogen und dem Kenntnisstand angemessen, sie vermeidet Trivialitäten, aber auch die Überbetonung des Formalismus. Sehr günstig ist es, daß in den meisten Arbeiten der experimentelle Hintergrund mit angegeben ist. Durch den institutsbezogenen Charakter der Publikation ist allerdings die Auswahl der Literaturzitate zu einseitig und repräsentiert nicht den jeweils aktuellen internationalen Stand.

Durch seine Vielfalt und durch seine problembezogene Betrachtungsweise bietet der Band Anregungen für alle, die an der mathematischen Durchdringung biologischer und medizinischer Prozesse arbeiten. *E. Winkler*

Periodic Table with Nuclides and Reference Data

K. Yoshihara, H. Kudo, T. Sekine. Berlin – Heidelberg – New York – Tokyo: Springer-Verlag 1985. X, 451 S., 47 Abb., 53 Tab.; geb., DM 198,–

In vielen Räumen von Kernforschungszentren und in Isotopenlaboratorien hängen an den Wänden die bunten Poster der Nuklide, die Nuklidkarten. Der Fachmann kann leicht die für ihn wissenswerten Angaben darauf finden: Atomgewicht, Massenzahl, Halbwertszeiten bei Radionukliden mit den entsprechenden Teilchenenergien, Wirkungsquerschnitte für Kernreaktionen, elektronische Konfigurationen u. ä. m. Die Autoren dieses Tabellenwerkes haben auch eine solche Nuklidkarte zusammengestellt; sie bieten aber außerdem eine erweiterte Fassung in Form eines Buches an.

Das Tabellenwerk ist in 8 Abschnitte eingeteilt, wobei nur einer der gewohnten Darstellung einer Nuklidkarte entspricht. Die folgenden Abschnitte sind speziell den Kerneigenschaften, den Kernreaktionen, der Wechselwirkung von Strahlung mit Materie und ihrer Detektion, den Radionukliden in der Chemie und in den Biowissenschaften, der Kernfusion und der Radioökologie einschließlich des Strahlenschutzes gewidmet. In einem Anhang werden gesondert die Naturkonstanten, die Umrechnungsfaktoren der verschiedenen Energieeinheiten, wichtige Formeln, die in Zusammenhang mit der Radioaktivität und der Statistik stehen, zusammengestellt.

Die Zahlenangaben werden i. allg. neuesten Veröffentlichungen entnommen, so beispielsweise den Angaben der IUPAC-Kommission für Atomgewichte und Isotopenhäufigkeiten aus dem Jahre 1981 oder der Nuklidkarte von *W. Seelmann-Eggebert* aus dem gleichen Jahr. Einige Unstimmigkeiten haben sich trotzdem eingeschlichen, wie beispielsweise die Halbwertszeit von ^{32}Si, die auf S. 14 mit 101 a angegeben wird, auf S. 416 aber mit 650 a; der Wert (104 ± 13) a wird heute meist verwendet. Die thermodynamischen Eigenschaften der Wasserstoffisotope werden sehr ausführlich behandelt; man wünschte sich an dieser Stelle Erweiterungen auch auf andere Isotope von Elementen, die häufig als Zahlenangaben benötigt werden, wie beispielsweise die Gleichgewichtskonstanten der Isotopenaustauschreaktionen $H_2O/HDO/D_2O$, $^{13}CO_2/^{12}HCO_3^-$, $^{13}CO_2/^{12}CO_3^{--}$, oder die thermodynamischen Eigenschaften von $H_2^{16}O/H_2^{17}O/H_2^{18}O$ u. a. Durch die Neudefinition des Meters auf der 17. Generalkonferenz für Maß und Gewicht in Paris am 20. 10. 1983 entfällt auch die Unsicherheitsangabe der Lichtgeschwindigkeit auf S. 441 des Anhanges.

Insgesamt ist das vorgelegte Tabellenwerk für den Praktiker eine sehr willkommene Zusammenstellung von wichtigen Daten und dürfte eine weite Verbreitung erwarten lassen, sei es in der Industrie oder in der Forschung oder auch in der Lehre.

H. Hübner

High-Dose Dosimetry

Vienna: IAEA 1985. 449 S., div. Abb. u. Tab. (Proc. Series, STI/PUB/671); karton. ÖS 880,–

Das internationale Symposium über Hochdosisdosimetrie wurde von der IAEA organisiert und fand vom 8.–12. Oktober 1984 in Wien statt. Über 70 Experten aus 30 Mitgliedsländern und 2 internationalen Organisationen nahmen an diesem ersten Symposium seiner Art teil. Es spiegelte das Interesse der Industrie an der Anwendung ionisierender Strahlung auf Gebieten wie der Strahlensterilisation medizinischer Instrumente, der Bestrahlung von Lebensmitteln, der Abfallbehandlung und der Plastherstellung wider.

In der ersten der drei Sitzungen wurden allgemeine Aspekte der Hochdosisdosimetrie an Großbestrahlungsanlagen in der Industrie und in der Forschung behandelt. *P. E. Vidal* (Association internationale d'irradiation industrielle) gab einen Überblick über Anforderungen an die Dosimetrie von Bestrahlungsanlagen. Dabei ging er auf die industrielle Routinedosimetrie, die Dosismessung unter extremen Umweltbedingungen, neue Meßmethoden unter Einbeziehung der Mikroelektronik und die Notwendigkeit von Dosimetervergleichen ein. Die übrigen Vorträge befaßten sich mit praktischen Lösungen und Erfahrungen an Bestrahlungsanlagen. Dabei standen Probleme der Dosimetrie an Gammaquellen in den Vorträgen von *W. Boeykens, L. Ghoos* (Mol, Belgien) und *H. M. Bär* et al. (Leipzig) sowie an Elektronenbeschleunigern in den Vorträgen von *P. P. Panta, Z. Bulhak* (Warschau) und *I. Kálmán* (Budapest) zur Diskussion. Die zweite Sitzung war neuen dosimetrischen Verfahren und Systemen gewidmet. In den meisten Vorträgen spielten Detektoren auf der Basis bekannter Nachweisprinzipien, die sich im Bereich geringerer Dosen bewährt haben, eine wesentliche Rolle. Spezielle aktive Systeme, wie z.B. *Geiger-Müller*-Zähler (*P. L. Lecuyer, P. M. Chaise*, Bollène, Frankreich), Ionisationskammern (*R. Tanka* et al., Takasaki, Japan) wurden ebenso wie passive Detektoren, wie z.B. *TLD* (*Dexi Jiang*, Bejing, VR China; *M. Osway, F. Golder*, Budapest) und chemische Detektoren (*A. Kovács* et al., Budapest; *D. Ražem* et al., Zagreb, *S. V. Klimov*, Moskau), vorgestellt. Besondere Aufmerksamkeit galt dem Alanine/ESR-Dosimeter in den Vorträgen von *D. F. Regulla, U. Deffner* (Neuherberg, BRD), *M. K. H. Schneider* et al. (Braunschweig) und *A. Bartolotta* et al. (Rom), mit dem die Dosisbestimmung im Bereich von 10 Gy bis 50 kGy mit einer Genauigkeit von ±5% möglich ist. In der dritten Sitzung wurden physikalische Aspekte der Dosisstandardisierung und Kalibrierung behandelt. Besondere Beachtung fand die Veränderung der optischen Eigenschaften bei Bestrahlung und deren physikalische Beeinflussung in verschiedenen Plasten, wie z.B. Perspex (*B. Whittaker* et al. und *K. M. Glover* et al., Harwell), Cellulosetriacetate (*P. Gehringer* et al., Seibersdorf, Österreich) und andere Plasten. Weitere Vorträge galten der Energieabhängigkeit verschiedener Dosimeter (*W. L. McLaughlin* et al., Gaithersburg, USA; *A. Miller*, Roskilde, Dänemark), der Standardisierung der Dosismessung (*W. L. McLaughlin*) und der Messung von Dosis- und Strahlprofilen (*R. Tanaka* et al.).

Allgemein wurde die Schlußfolgerung gezogen, daß eine weitere Verbesserung und Vervollkommnung der existierenden dosimetrischen Verfahren notwendig und zweckmäßig ist. Der vorliegende Tagungsband enthält wichtige Informationen zum gegenwärtigen Stand auf dem Gebiet der Dosimetrie im Bereich hoher Dosen, die in erster Linie den Fachmann für Dosimetrie und Strahlenschutz ansprechen. Aber auch für den Physiker und Chemiker dürften verschiedene Effekte in Stoffen bei hohen Bestrahlungsdosen von Interesse sein.

G. Pretzsch

Lexikon der Kybernetik

Im Auftrag des Zentralinstituts für Kybernetik und Informationsprozesse der Akademie der Wissenschaften der DDR herausgegeben
von GÜNTER LAUX

(Übersetzung aus dem Russischen)

Band I: A—E

1980. 755 Seiten — 120 Abbildungen — 19 Tabellen — 14,5 cm × 21,5 cm — Lederin
DDR 70,— M; Ausland 70,— DM
Bestell-Nr. 762 1920
Bestellwort: Lex. Kyb. A—E (6297/1)

Band II: F—L

1981. 642 Seiten — 135 Abbildungen — 15 Tabellen — 14,5 cm × 21,5 cm — Lederin
DDR 60,— M; Ausland 60,— DM
Bestell-Nr. 762 1939
Bestellwort: Lex. Kyb. F—L (6297/2)

Band III: M—R

1981. 704 Seiten — 123 Abbildungen — 12 Tabellen — 14,5 cm × 21,5 cm — Lederin
DDR 65,— M; Ausland 65,— DM
Bestell-Nr. 762 6019
Bestellwort: Lex. Kyb. M—R (6297/3)

Band IV: S—Z

1981. 589 Seiten — 97 Abbildungen — 16 Tabellen — 14,5 cm × 21,5 cm — Lederin
DDR 55,— M; Ausland 55,— DM
Bestell-Nr. 762 6027
Bestellwort: Lex. Kyb. S—Z (6297/4)

Neu:
Band V: Ergänzungsband

1985. 560 Seiten — 97 Abbildungen — 14 Tabellen — mit Abkürzungs- und mehrsprachigem Schlagwortverzeichnis — 14,5 cm × 21,5 cm — Lederin DDR 60,— M; Ausland 60,— DM
Bestell-Nr. 763 407 8
Bestellwort: Lex. Kyb. (6297/E)

Bitte richten Sie Ihre Bestellungen an eine Buchhandlung

AKADEMIE VERLAG
THEORIA CUM PRAXI
BERLIN
Deutsche Demokratische Republik
1946-1986
DDR-1086 Berlin, Leipziger Straße 3-4 · Postfach 1233

ISOTOPENPRAXIS

Band 22 1986 Heft 2

		Contents	Inhalt
		Articles	**Originalarbeiten**
St. Szpilowski, J. St. Michalik, A. Owczarczyk	41	►Prediction of Removal Efficiency of Rectangular Settling Tank. Part I. Application of Radiotracer Method in Tank Testing and Laboratory-Scale Verification of the Mathematical Model of Sediment Transport	Voraussage der Abtrennungseffektivität von rechteckigen Abscheidetanks. Teil I: Die Anwendung der Radiotracermethode bei der Testung von Tanks im Labor — bei der Prüfung der Maßstabsübertragung des mathematischen Modells auf den Sedimenttransport
St. Szpilowski, J. St. Michalik, A. Owczarczyk	46	►Prediction of Removal Efficiency of Rectangular Settling Tank. Part II. Investigation of Commercial Scale Settlers	Voraussage der Abtrennungseffektivität von rechteckigen Abscheidetanks. Teil II. Untersuchungen an einem Abscheider im technischen Maßstab
K. G. Kruszewski, St. Michalik	49	►Identification of the Molten Glass Flow Process in Tank Furnaces	Identifizierung des Fließprozesses von geschmolzenem Glas in Wannenöfen
O. P. D. Noronha	53	►Solvent Extraction Technology of ^{99}Mo—^{99m}Tc Generator System. 1. An Indian Experience Process Design Considerations	Die Technologie der Lösungsmittelextraktion bei einem ^{99}Mo—^{99m}Tc-Generatorsystem. 1. Eine indische Erfahrung: Überlegungen zur Prozeßauslegung
O. P. D. Noronha	58	►Solvent Extraction Technology of ^{99}Mo—^{99m}Tc Generator System. 2. Radiation Dose and Personnel Exposure Considerations	Technologie der Lösungsmittelextraktion bei ^{99}Mo—^{99m}Tc-Generatorsystemen. 2. Strahlungsdosis- und Personendosis-Betrachtungen
A. Pinkowski, W. Hübner, H.-J. Ulrich	63	►Investigation of Mechanically Treated Iron Electrodes by Means of Autoradiography and *Kossel* Technique	Untersuchung mechanisch behandelter Eisenelektroden mit Hilfe der Autoradiographie und der *Kossel*-Technik
D. Müller, P. Morgenstern	65	Determination of the O/U-Ratio of UO_{2+x} by X-Ray Fluorescent Analysis	►Bestimmung des O/U-Verhältnisses von UO_{2+x} mit röntgenradiometrischen Mitteln
G. Krüger	68	Rapid Determination of Arsenic in Ores and Dressing Products by Radionuclide X-Ray Fluorescence Analysis	►Schnelle Bestimmung von Arsen in Erzen und Aufbereitungsprodukten mit der Radionuklid-Röntgenfluoreszenzanalyse
	70	**New Books**	**Buchbesprechungen**

►Language of Publikation ►Publikationssprache

Isotopenpraxis is indexed in Current Contents/PC & ES, Science Citation Index, and INIS Atomindex

FSC
www.fsc.org
MIX
Papier | Fördert gute Waldnutzung
FSC® C083411

Zeitfracht Medien GmbH
Ferdinand-Jühlke-Straße 7
99095 Erfurt, Deutschland
produktsicherheit@kolibri360.de